KINETICS IN MATERIALS SCIENCE AND ENGINEERING

KINETICS IN MATERIALS SCIENCE AND ENGINEERING

Editor

Khamisi Zaman

scitus
academics

Kinetics in Materials Science and Engineering
Edited by **Khamisi Zaman**

ISBN: 978-1-68117-721-2
Library of Congress Control Number: 2016958817

SCITUS Academics LLC,
616, Corporate Way, Suite 2, 4766,
Valley Cottage, NY 10989

www.scitusacademics.com

Notice

Reasonable efforts have been made to publish reliable data and views articulated in the chapters are those of the individual contributors, and not necessarily those of the editors or publishers. Editors or publishers are not responsible for the accuracy of the information in the published chapters or consequences of their use. The publisher believes no responsibility for any damage or grievance to the persons or property arising out of the use of any materials, instructions, methods or thoughts in the book. The editors and the publisher have attempted to trace the copyright holders of all material reproduced in this publication and apologize to copyright holders if permission has not been obtained. If any copyright holder has not been acknowledged, please write to us so we may rectify.

Printed in United States of America on Acid Free Paper

PREFACE

Materials are valuable to human civilization due to the advantages of manipulating their properties, e.g. electrical conductivity, dielectric constant, magnetization, optical transmission, toughness and strength. The science of materials is a broad area and can be regarded as an interdisciplinary area. The studies of the structure and properties of each material, the development of new types of material and the distortion of the properties of a material to meet the needs of a new application are included. Materials are actually more important than we realize in our culture. Transport, housing, communication of clothing, reaction and food production and almost every segment of our everyday lives are shaped by materials. Materials have contributed significantly to the development of a variety of technologies, such as medicine & health, information & communication, national security & space, transport, structural materials, arts & literature, textiles, personal hygiene, agriculture & food & environmental science. These interdisciplinary relationships between the material sciences and several other areas in the development and deployment of new materials should be well understood.

Kinetics in Materials Science and Engineering presents a compilation on the current state of the material know-how: synthesis, characterization, properties and applications. The majority of chapters include a review of current knowledge and state-of-the-art results. The book demonstrates the important role of selecting materials in the design and manufacturing mechanisms of new products. The engineering and design mechanisms cannot be separated from either the material design, which is increasingly assisted by computers or the modern technological design of the most appropriate production processes. The review of the multi-millennium long history of human civilization shows that the modest increase in living and production levels is more often linked to the creation of new material groups with properties that are better adapted to the real needs of customers who are becoming more sophisticated almost every day and the launch of technological processes. This book is therefore an essential reference source for scientists in the fields of energy, physics, chemistry and materials. This book is also an appropriate reading material for graduates.

TABLE OF CONTENTS

1

ELECTRICAL, THERMAL AND MECHANICAL PROPERTIES OF CNT TREATED PREPREG CFRP COMPOSITES

Yelda Akcin[1], Sukru Karakaya[2], Omer Soykasap[3]

[1]Faculty of Technology, Metallurgy and Materials Engineering, Afyon Kocatepe University, Afyonkarahisar, Turkey

[2]Faculty of Technology, Mechanical Engineering, Afyon Kocatepe University, Afyonkarahisar, Turkey

[3]Faculty of Engineering, Materials Science and Engineering, Afyon Kocatepe University, Afyonkarahisar, Turkey

ABSTRACT

This study is part of Smart Intelligent Aircraft Structures (SARISTU) project, which aims considerable improvements in aircraft damage tolerance, electrical conductivity and weight reduction besides producibility in industrial scale. In this study, the effect of multiwalled carbon nanotube reinforcement on electrical, thermal and mechanical properties of T800/M21 carbon fibre reinforced plastic is studied experimentally. T800/M21 is a commercial prepreg carbon fibre/epoxy composite material considered for CNT treatment by means of CNT-doped thermoplastic-based dry powder. The CNTs are deposited on top of prepreg material uniformly using a controlled spraying machine selecting the best state-of-the art and innovative performing technology from the candidate technologies within the project. The electrical conductivity of the composite material with/without CNT is measured in longitudinal, transverse and thickness directions. The changes occurring in the electrical conductivity of the composite materials are investigated. In order to investigate thermal behaviour of the composite materials, differential scanning calorimetry and thermogravimetric analyses are performed. Detailed thermal analysis is conducted for with/without carbon nanotube reinforced material to obtain the thermal conductivity, specific heat and thermal expansion coefficient of the material. Finally, the effect of carbon nanotube reinforcement on mechanical behaviours is studied by tensile, bending and shear tests.

Keywords: Carbon Nanotubes, Electrical Properties, Thermal Properties, Mechanical Properties, CFRP

1. INTRODUCTION

Fibre reinforced polymer composite (CFRP) materials provide a significant mass saving in production due to their high strength/mass and high stiffness/mass ratios compared with metals. Lighter aircraft with multifunctional materials such as carbon nanotube (CNT) doped CFRP provides energy and performance efficiency; hence research on implementation of CNT doping on commercially available CFRP composites in the aerospace industry has focused on easy and environment-friendly production in an industrial scale in recent years [1] . CNT has extraordinary electrical, thermal and mechanical properties making them potentially attractive materials for use in aerospace structures in order to enhance the material properties. Epoxy polymers are the most commonly used matrix because of their mechanical properties along with their compatibility with a large variety of fibres. This material ensures superior properties of high modulus, high thermal and chemical resistance, and low creep for most aircraft structural applications. However, epoxy resin is brittle and allows the formation of cracks between fibre and matrix in the composite material subject to certain loading conditions [2] , especially during liquid moulding process due to viscosity of epoxy [1] .

Electrical conductivity of epoxy is very low and even if it is used as a matrix for conductive carbon fibre, the produced CFRP composite material cannot reach required electrical conductivity for electrical structure network of airframes to prevent failure especially in thickness direction [3] . The electrical functions of airframes such as electro static discharge, electrical bond, electromagnetic shielding, and lightning strike protection are closely associated with electrical conductivity of the materials. For example, a possible structural and electronic damage may occur due to lightning strike when the airframe is not conductive enough. In order to protect CFRP structures of aircraft, the current state of the art metal wire mesh or expanded aluminium and copper materials are incorporated into the surface of these composite structures so that the lightning strike energy is dissipated over the surface and hence it prevents damage of the composite material beneath. On the other hand, adding mass increases the weight of aircraft structures which degrades performance efficiency of the aircraft. If a conductive material can be produced directly without the need of coating the surface after production, the installation cost of electrical structure network decreases. In literature, there are many studies on enhancing the electrical conductivity and mechanical properties of epoxy by adding fillers such as short carbon fibre, carbon nanofiber, carbon nanotube, etc. As a common result of these studies, all the fillers improve the electrical conductivity and mechanical behaviours of epoxy [4] - [8] .

In recent years, CNTs are considered the most popular materials as fillers for CFRP materials. Because, they have high potential to improve electrical conductivity, mechanical and thermal properties of polymer material simultaneously without causing to weight increment [9] . To improve damage tolerance of composite materials, the usage of thermoplastic veils as interleaf materials and thermoplastic carriers for implication of nanoparticles are preferred methods. It has been reported that remarkable improvements are obtained in the fracture toughness of composite materials produced with these methods [10] [11] .

The thermal behaviours of polymer used in CFRP are as important as mechanical and electrical properties since aircraft structures are exposed to thermal loads as well. Resin melting, vaporization and ply delamination may occur due to the poor thermal conductivity or electrical resistance of composite materials during lightning strike or electrical discharge. The studies on the effect of CNT doping on thermal conductivity exhibit that CNT doping produces slight enhancement in the thermal conductivity as opposite to electrical conductivity [12] [13] . According to many studies on CNT doped CFRP composites, it is seen that the improvement and good performance is related to dispersion of CNT in the matrix which is the main issue of the manufacture method used [14] - [17] .

Production of CNT doped carbon fibre/epoxy composite with conventional liquid injection technologies causes some problems. During doping proses, CNTs are mixed and interact with the resin, causing an increase its viscosity. High viscosity obstructs the flow of the resin and resin impregnation of fibres decreases accordingly. The viscosity can be reduced by increasing the process temperatures but in this case CNTs show tendency to agglomerate. On the other hand, the CNTs collected around the carbon fibres during the infusion process can lead to formation of deflective layers [1] . CNTs with good distribution in the matrix improve many properties of CFRP composite materials without increasing structural weight whereas they cause degradation of electrical conductivity, mechanical and thermal behaviours when they are agglomerated. This study is as part of Smart Intelligent Aircraft Structures (SARISTU) project, which aims considerable improvements in aircraft damage tolerance, electrical conductivity and weight reduction. Several state-of-the art and innovative solutions such as CNT doped veil, CNT doped binder, CNT treated prepreg etc. were studied within SARISTU to enhance the damage tolerance and electrical conductivity of composite materials. Besides these aims, producibility of composite material in industrial scale was taken in consideration as an important effect. As the results of studies, the two best performing technologies, 1) thermoplastic veils and 2) CNT treated prepreg with spraying method are selected to transfer from the laboratory scale to demonstrator production and accordingly to be studied at an industrial scale. For inclusion of CNT to CFRP composite, production with the thermoplastic veil is suitable for liquid composite moulding while the spraying method is suitable for prepreg [1] . In this study, the effect of CNT addition with spraying method on electrical, mechanical and thermal behaviour of unidirectional carbon fibre/epoxy prepreg composite material is investigated experimentally. T800/M21 is a commercial prepreg carbon fibre/epoxy composite material, which is considered CNT treatment by means of CNT-doped thermoplastic-based dry powder. For the production of CNT doped carbon fibre/epoxy composite materials, powder deposition is used to enable production on an industrial scale and resolves agglomeration problems during doping. Section 2 describes the experiments of thermal, mechanical and electrical studies. The electrical conductivity of neat and CNT doped composite material is determined at room temperature in longitudinal, transverse, and thickness directions. Moreover, the electrical conductivity is analysed at lower and higher temperatures in the thickness direction. For the thermal behaviour, Differential Scanning Calorimetry (DSC) and Thermogravimetric (TG) analyses are performed for the measurements of specific heat capacity, coefficient of thermal conductivity and coefficient of thermal expansion. Mechanical behaviour of CNT doped material is investigated by tensile, three point bending and shear tests. Next the experimental results are discussed in Section 3. Main conclusions of the paper are presented in the final section.

2. EXPERIMENTAL STUDIES

The neat material used in the study is T800/M21 carbon fibre/epoxy prepreg, which is commercially available, aerospace qualified unidirectional material with an areal density of 198 g/m^2. Production of CNT-treated CFRP prepreg and curing the CFRP material for the specimens are carried out by the project partners as described in what follows. The NC7000 multi walled carbon nanotube (MWCNT) is produced by catalytic chemical vapour decomposition by the project partner Nanocyl SA, Belgium. MWCNT NC7000 has the following properties: an average diameter of 9.5 nm, an average length of 1.5 μm, carbon purity of 90%, metal oxide of 10%, surface area of 250 - 300 m^2/g. Nanocyl aimed at preventing agglomeration of CNTs when added to resin and manufacturing ease for prepreg materials, and achieved a good dispersion of CNTs in thermoplastics to produce compound PLASTICYLTM using industrial twin-screw extruder in pellet form. These pellets can be produced via a wide range of thermoplastic resin. For this study, they are produced from Griltex D1330A Polyamide. The reason of selecting polyamide resin is its melting temperature ranging between 125°C - 135°C in order to melt the polyamide resin during the curing cycle and to migrate in CFRP. CNT doped prepreg materials are produced by the other project partner University of Patras, Greece. CNT-doped thermoplastic granules are ground into powder grain size in the range 100 to 250 μm. Then, these powders are deposited on top of T800/M21 carbon/epoxy prepreg material as 2 g/m^2 in a controlled spray machine developed specifically for this purpose. The process is customized to deliver nano-treated prepregs and prevents the formation of agglomeration. Finally both neat and CNT doped T800/M21 carbon fibre/epoxy prepreg materials are supplied and laminates are produced by the other project partner Turkish Aerospace Industries Inc., Turkey. The materials are cured by autoclave with curing cycle of 2 hours under 8 bar pressure at 180°C temperature, next cut into required specimen sizes. Optic microscope and scanning electron microscope (SEM) images taken from the cross sections of the specimens are given in Figures 1-3 for both neat and CNT doped composite material. It is seen that there is no agglomeration of CNTs, which are dispersed uniformly in epoxy resin. The spraying method ensures a good dispersion and manufacture ease. Thermal, mechanical and electrical conductivity tests are described in the following subsections.

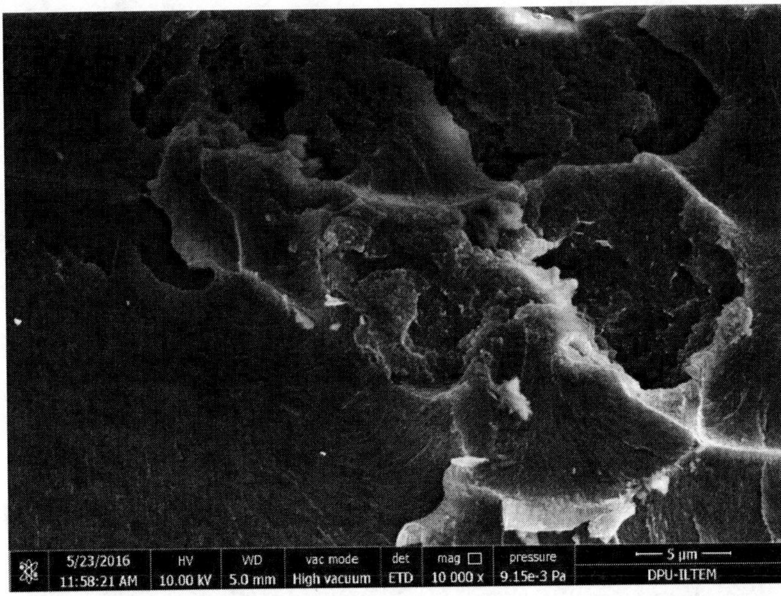

(a)

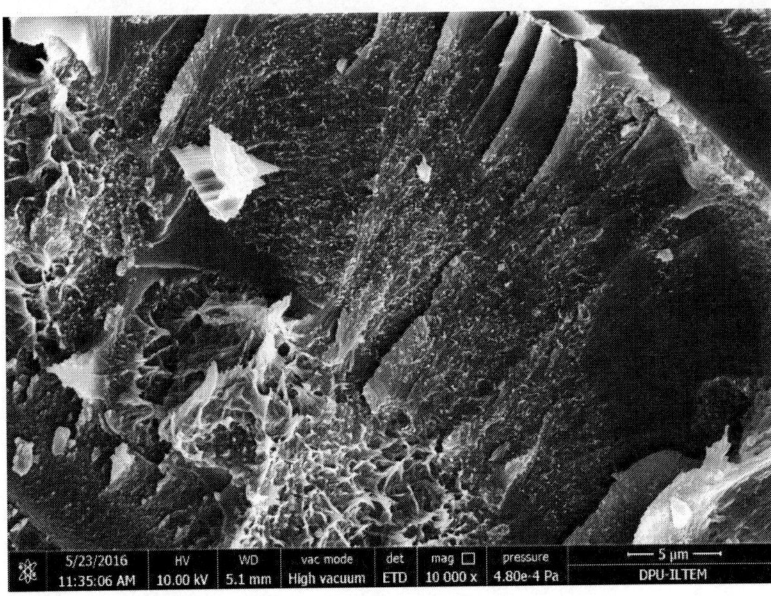

(b)

Figure 1.SEM images of the material. (a) Neat carbon/epoxy composite specimen; (b) CNT doped carbon/epoxy composite specimen.

(a)

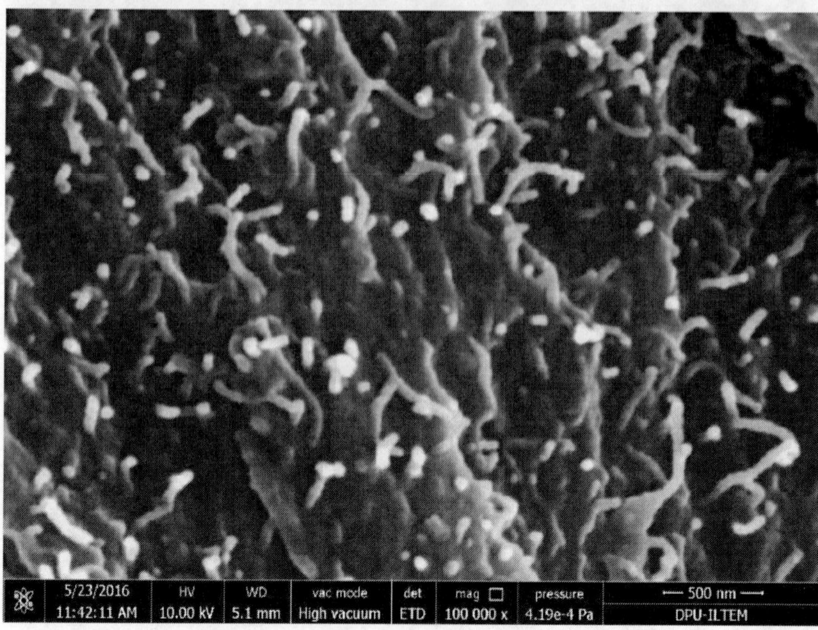

(b)

Figure 2. High-magnification images of CNT doped carbon/epoxy composite specimen. (a) 25,000×; (b) 10,000×.

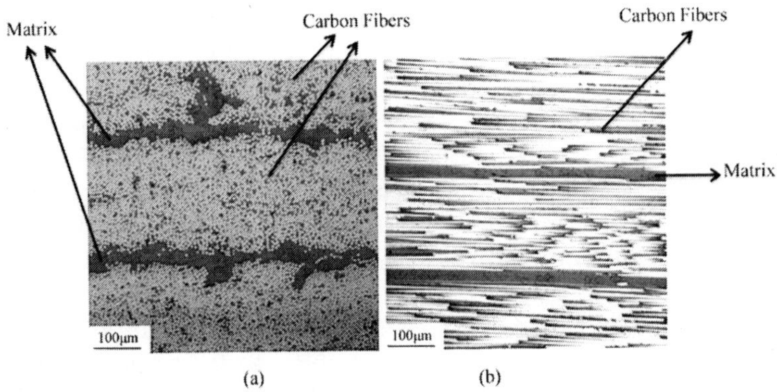

Figure 3. Optic microscope images of the material taken from (a) a longitudinal section, (b) a transverse section.

2.1. Thermal Tests

Thermal behaviour of the materials is investigated by several analyses and measurements including DSC, TG, specific heat capacity, thermal conductivity, and thermal expansion. Coefficient of thermal conductivity of neat and doped specimens is determined at the room temperature in the longitudinal, transverse and thickness direction with C-THERM/TCi thermal conductivity device. Surface of specimens with dimensions of 50 × 50 × 4 mm (length × width × thickness) is prepared by sanding before the measurement. Next, DSC, TG and specific heat capacity analyses are performed with NETZSCH Differential Scanning Calorimetry. The specimens with dimensions of 5 × 5 × 4 mm (length × width × thickness) are tested at the temperatures from 24°C to 110°C with a heating rate of 20°C/min at atmospheric conditions. Finally, to determine the dimensional changes of the composite materials due to temperature effect, the coefficient of thermal expansion of 25 × 5 × 4 mm (length × width × thickness) specimens is measured in the longitudinal and transverse directions with NETZSCH 402 CD Dilatometer between 30°C and 180°C.

2.2. Mechanical Tests

The mechanical properties of neat and doped T800/M21 carbon/epoxy specimens are investigated via tensile, bending and shear tests. First, the tensile tests are performed with Instron 8801 tensile testing machine with video extensometer. The tensile tests are performed according to ASTM D 3039M standard in the longitudinal and transverse directions. The specimens with a dimension of 250 × 20 × 2 mm (length × width × thickness) are prepared with layups (0)n, (90)n. The metal tabs are affixed at the end of the specimens to prevent premature failure; and the tests are carried out with a speed of 2 mm/min. Tensile strength, elastic modulus, Poisson's ratio and failure strain are obtained as results of tensile tests. Next, three point bending tests are applied to neat and doped specimens according to ASTM D790 standard. The specimens with dimensions of 120 × 10 × 2 mm (length × width × thickness) are tested with a head speed of 5 mm/min. The bending modulus of specimens is obtained. Last, the shear tests are performed according to ASTM D3518 standard to obtain shear modulus of specimens. For these tests, the specimens with dimensions of 100 × 25 × 2 mm (length × width × thickness) and layup (±45)n are prepared and the tensile test is performed. For all mechanical tests, six specimens are prepared and tested and then average is taken.

2.3. Electrical Conductivity Tests

A test apparatus is established to measure electrical conductivity of specimens at room temperature. The schematic view of established test apparatus is given in Figure 4. There are two Teflon parts mounted on a rod and two copper parts mounted on the Teflon parts for the testing apparatus. One of the Teflon parts is fixed whereas the other is movable horizontally so that the measurement length is adjustable. The apparatus allows the measurement of electrical conductivity of specimens in longitudinal, transverse and thickness directions. In order to reduce the contact resistance between the specimen and the probes, the surface of specimens are coated with silver paint, and the measurements are performed with four-probe-method in order to eliminate any inference of probe resistance [18] . First, the specimens are placed between the copper plates, and full contact is ensured by applying force, then the electrical conductivity of specimens is measured with HM8118 LCR-Bridge at room temperature. Three neat and doped specimens are prepared for each direction, the measurements are repeated five times and the results are averaged. In order to further investigate the change of electrical conductivity with temperature in the thickness direction, the electrical conductivity of neat and CNT doped materials are measured with Novacontrol Concept 50 Dielectric and Impedance Spectrometer at 1 kHz frequency and between the temperatures −100°C and 160°C.

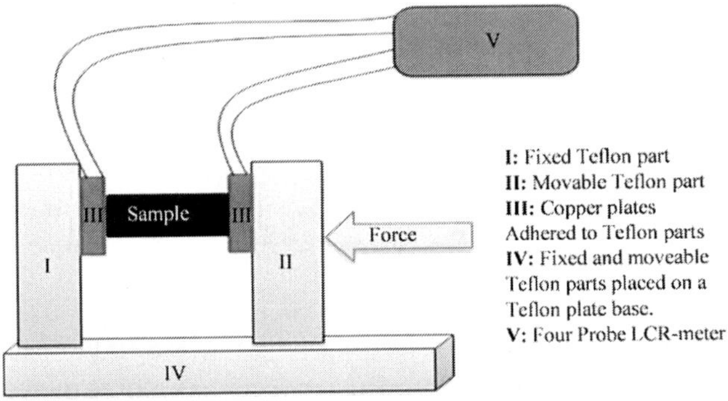

Figure 4. Test apparatus to determine the electrical conductivity of specimens at the room temperature.

Table 1. Thermal conductivity of neat and CNT treated T800/M21.

Direction	Thermal Conductivity Coefficient, neat (W/mK)	Thermal Conductivity Coefficient, CNT treated (W/mK)
Longitudinal	4.60	4.53
Transverse	0.80	0.71
Thickness	0.69	0.67

3. RESULTS

3.1. Thermal Tests

The thermal conductivity of neat and doped specimens is measured in longitudinal, transverse and through the thickness and the results are given in Table 1. The thermal conductivity of the doped specimens decreases slightly in all directions. While the coefficient of thermal conductivity of specimens in longitudinal is very high, it is relatively lower in the transverse and thickness direction. According to quantum mechanics, thermal conduction in carbon materials is carried by phonons in which the atoms or molecules uniformly oscillates with vibration modes. Although CNTs have high thermal conductivity, and hence it was expected a higher thermal conductivity due to CNT treatment according to the rule of mixtures, there is a decrease in the thermal conductivity in three directions due to CNT treatment. However thermal conduction is significantly affected by scattering any structural defects, discontinuities, and layer interactions. According to the studies, discontinuities in polymer lead to reduction of mean free oscillation of polymer molecules and electron scattering [19] . So, it is concluded that CNT treatment does not increase conduction of the carbon fibre/epoxy system. The thermal conductance and electrical resistance can cause resin melting, vaporization and ply delamination, reduction of thermal conductivity is a preferred feature for carbon fibre/polymer composites [20] .

DSC diagrams of the neat and CNT doped specimen are given in Figure 5 and Figure 6, respectively. The Figure 6 is zoomed partition of Figure 5 in order to see the glass transition temperature closely. The first peak is seen around 200°C and this corresponds to the glass transition temperature. After this temperature, the system shows two exothermic reactions with heat output during which degradation reactions of epoxy matrix occur. First peak indicates the decomposition of lower molecular weight components that have thermally weak bonds and the other one is the degradation of higher molecular weight components formed during curing as discussed in the references [21] - [24] . The phase changes of the neat and doped materials are similar according to the diagrams. The phase change temperatures are compared in Table 2 in more details, in which T_{onset} is the starting temperature of reaction; T_g is glass transition temperature; T_{finish} is the finishing temperature of reaction and T_p is peak point of the reaction. While the glass transition temperature of neat specimen is 207°C, the glass transition temperature of CNT doped specimen is 212°C with a 5°C increment. The same increment is also seen in the decomposition reactions for CNT doped specimen. According to the studies on polymeric materials, addition of filler materials may prevent movement of the molecules and diffusion; thereby this situation leads to an increase in the glass transition and decomposition temperatures as discussed in the references [19] [25] [26] .

The peaks labelled as the glass transition temperature and the decomposition reaction in DSC diagrams is also confirmed with TG analyses in Figures 7-9. Figure 7 and Figure 8 show the combination of DSC and TG curves, the first degradation reaction in DSC diagrams shows itself with a sudden mass loss approximately between 250°C - 450°C and the second degradation reaction starts after 450°C in the TG diagrams.

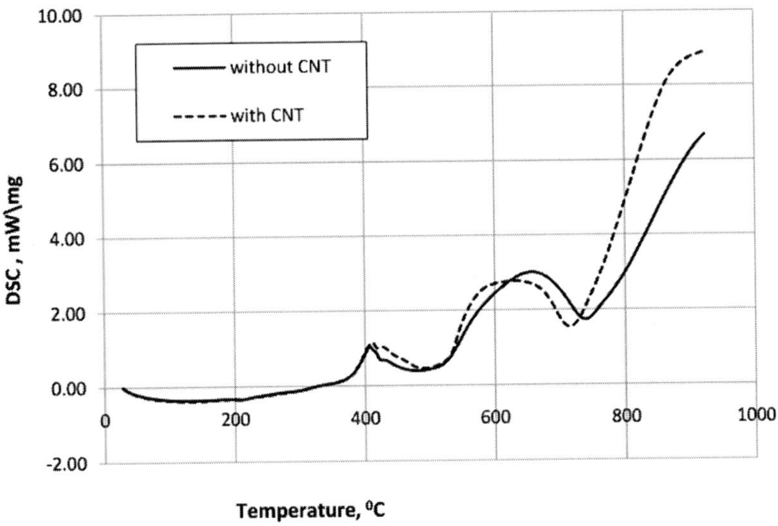

Figure 5. DSC thermograms of neat and CNT doped T800/M21.

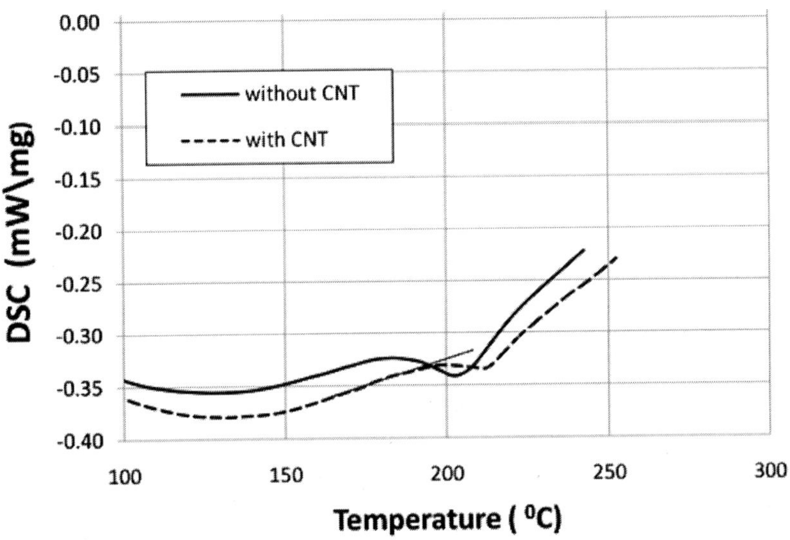

Figure 6. Glass transition temperatures of neat and CNT doped T800/M21.

Table 2. DSC results of neat T800/M21 and CNT doped T800/M21.

Specimens	1. Peak			2. Peak			3. Peak		
	T_{onset} (°C)	T_g (°C)	T_{finish} (°C)	T_{onset} (°C)	T_p (°C)	T_{finish} (°C)	T_{onset} (°C)	T_p (°C)	T_{finish} (°C)
Neat T800/M21	177	202	222	320	407	487	492	657	742
CNT doped T800/M21	197	212	227	327	412	492	502	642	722

When the TG diagrams of neat and CNT doped specimens are compared, the mass losses are quite close for both specimens until the start of first degradation reaction as seen in Figure 9. When the degradations begin, more mass loss observed in CNT doped specimens than the neat specimens (41% mass loss in neat specimens whereas 48% mass loss in doped specimens at 830°C). More mass loss is seen in CNT doped specimens because CNTs combust around 600°C in air and get away from system at atmospheric conditions according to the studies [27] [28] .

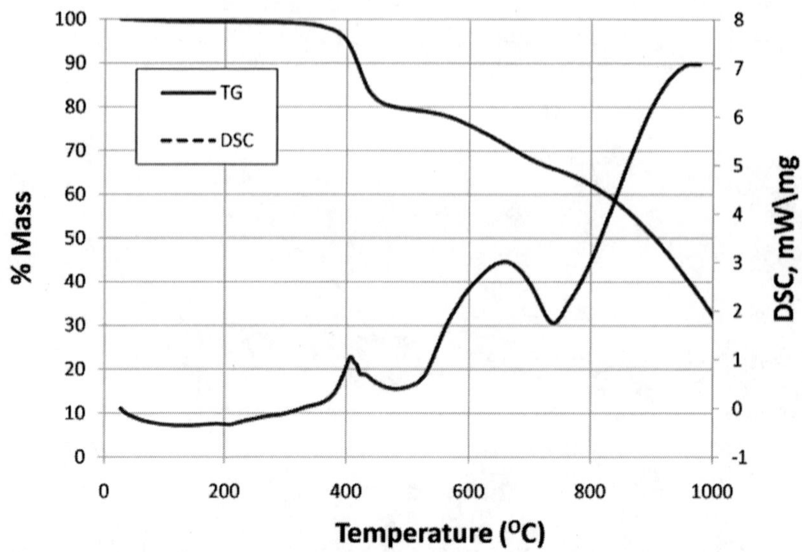

Figure 7. TG thermograms of neat T800/M21.

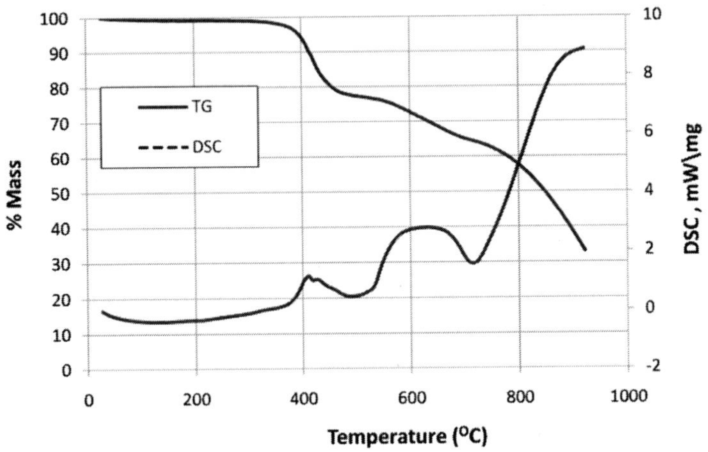

Figure 8. TG thermograms of CNT doped T800/M21.

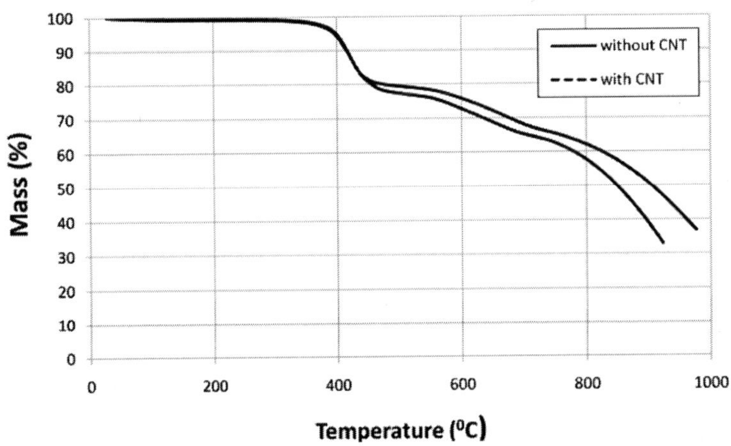

Figure 9. Comparative TG thermograms of neat and CNT doped T800/M21.

The specific heat capacity increases with temperature for both neat and doped specimens as given in Figure 10. However, the value of the doped specimens is always lower than that of the neat specimens. The specific heat capacity is obtained as 0.86 J/gK for the neat material and 0.80 J/gK for CNT doped material with a 7% reduction under constant pressure at 50°C. The specific heat capacity is described as the amount of heat necessary to raise the temperature of a certain mass by 1 degree Kelvin. If evaluated in atomic size, heat capacity is stored energy as atomic vibrations in materials. Depending on this, CNTs situated in the

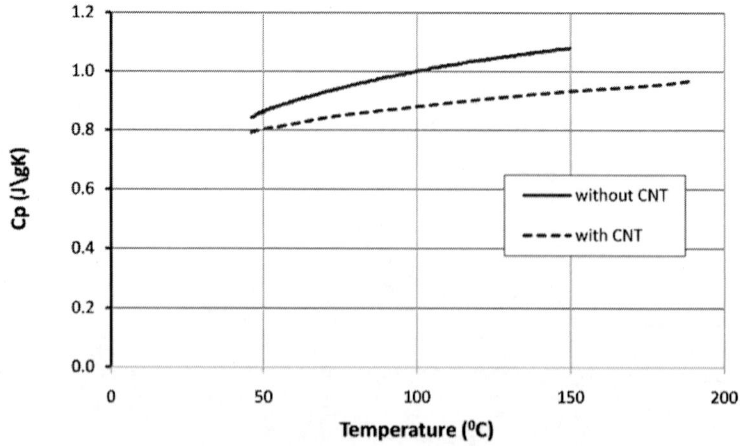

Figure 10. Specific heat capacity of neat T800/M21 and CNT doped T800/M21 specimens.

epoxy hinder the movement of polymer molecules and cause a decrease in total kinetic (thermal) energy of the system. Thus, the specific heat capacity of CNT doped specimens is lower than that of the neat specimens.

The coefficient of thermal expansion (CTE) is an important property for polymers used in engineering applications. For good dimensional stability, a low CTE is demanded. Aerospace industry is a pioneer in the use of composite materials comprised of fibres with a low CTE and resin with high CTE. The aim is to reduce the total thermal deformation for good thermal stability [29] . The CTE of M21 epoxy matrix is 55×10^{-6}/°C whereas CTE of T800 carbon fibre is -0.56×10^{-6}/°C [30] [31] . During heating, the matrix tends to expand whereas the carbon fibre tends to shrink with increasing temperature. According to Figure 11, the CTE of neat T800/M21 composite material is positive and low in longitudinal direction over the range of 20°C and 180°C. On the other hand, CTE of CNT doped material is almost zero, negative up to 140°C and lower in magnitude than that of the neat material. Behaviour of the carbon fibre dominates the expansion in longitudinal direction. Further obstructive effect of CNTs on polymer chains possibly leads to negative effect on expansion of epoxy in CNT doped system, and the composite material is shrunken with a reduction of the CTE. On the other hand, the matrix plays an important role in transverse direction for CTE. CTEs are higher in transverse direction compared to the longitudinal direction as expected. The measured CTEs of the specimens increase with the temperature between 30°C and 180°C as given in Figure 12. CNT doped specimens are expanded more than the neat specimens in transverse direction, the effect is much more significant at lower temperatures. In literature, there are some studies showing that addition of CNT in polymer matrix causes an increase in stability [32] [33] . Even if the result for the longitudinal direction support this studies, the same effect is not seen in transverse direction. It is thought that this situation is related to the production method and weight percentage of the CNTs.

3.2. Mechanical Tests

The tensile strength, elastic modulus, Poisson's ratio and failure strain are measured in fibre and transverse directions. The results are summarized in Table 3 for the neat and in Table 4 for CNT doped materials. In the tables, w denotes the specimen width, t denotes the specimen thickness, and E denotes the elastic modulus of specimens.

According to Table 3 and Table 4, addition of CNT in carbon/epoxy composite system increases the tensile strength by 7%, the elastic modulus by 4%, and the failure strain by 16% in the longitudinal direction. In the transverse direction, the tensile strength increases by 21%, the elastic modulus increases by 21% and the failure strain increases by 9%. CNTs show the greatest effect on the failure strain in longitudinal direction and on the tensile strength in transverse direction.

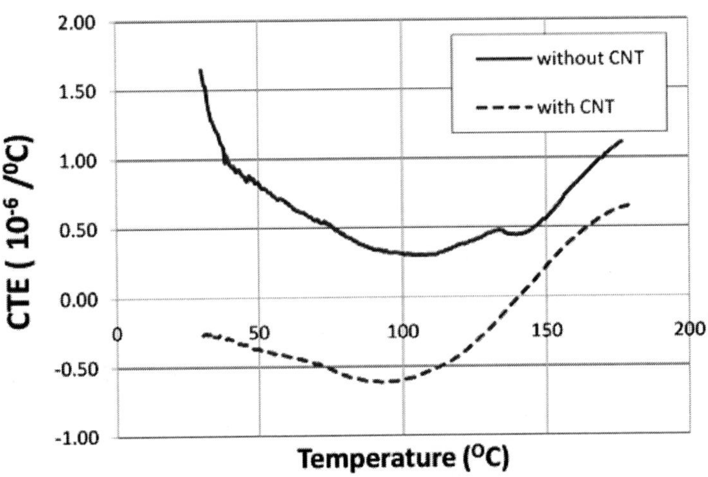

Figure 11. CTE of neat and CNT doped T800/M21 specimens in the longitudinal direction.

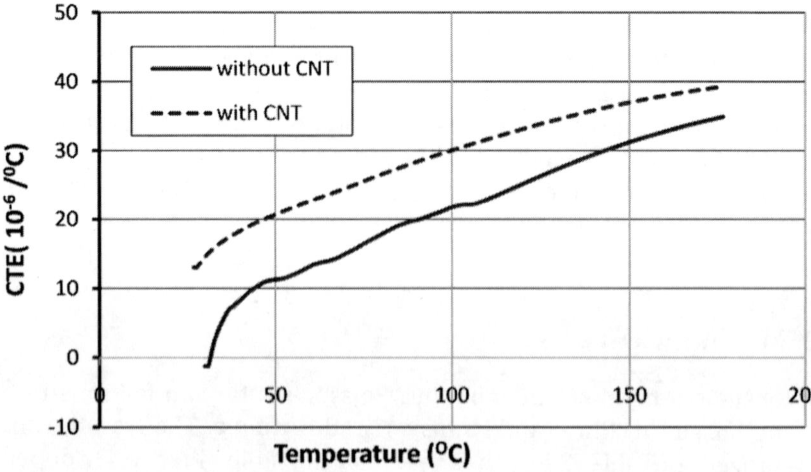

Figure 12. CTE of neat and CNT doped T800/M21 specimens in the transverse direction.

Table 3. Tensile test results of neat T800/M21.

Layup	w (mm)	t (mm)	E (GPa)	Poison's ratio	Tensile strength (MPa)	Failure strain (%)
$(0)_n$	20.4	2.0	158.94	0.395	1371.6	0.87
$(90)_n$	20.4	2.0	8.65	-	30.97	0.32

Table 4. Tensile test results of CNT doped T800/M21.

Layup	w (mm)	t (mm)	E (GPa)	Poison's ratio	Tensile strength (MPa)	Failure strain (%)
$(0)_n$	20.4	2.0	165.15	0.38	1476.25	1.01
$(90)_n$	20.4	2.0	9.38	-	37.48	0.35

The three point bending test is applied to determine the effect of CNT treatment on the flexural properties of carbon/epoxy specimens. The obtained results for the neat and doped specimens are given in Table 5, where s is the distance between supports, Ef is bending modulus. The addition of CNTs increases the bending modulus by 11% approximately for both fibre and transverse directions.

The shear test is performed to obtain shear modulus, G12. The test results are given in Table 6 for both materials. According to the results, the shear modulus increases by 12% in the carbon/epoxy system with CNT addition.

Considering all the mechanical test results, CNT addition improves all mechanical properties. In fibre/epoxy composite systems, fibre-matrix interface is one of the most important resources that cause crack initiation during loading, because the formation of cracks starts in the fibre-matrix interface generally. Presence of CNTs in carbon/ epoxy composite material prevents the formation and propagation of the cracks by affecting load distribution. Thus, it improves the mechanical properties of carbon/epoxy composite materials as discussed in the references [14] [15] . It should be noted that the studies realized within project of SARISTU to improve the damage tolerance of carbon fibre/epoxy composite materials show that, fracture toughness for mode I, (Gic) of

Table 5. Bending test results of neat and CNT doped T800/M21 specimens.

	Neat T800/M21				CNT doped T800/M21			
Layup	w (mm)	t (mm)	s (mm)	Ef (GPa)	w (mm)	t (mm)	S (mm)	Ef (GPa)
$(0)_n$	10.2	2.0	60	128.56	10.2	2.0	60	142.81
$(90)_n$	10.2	2.0	60	6.47	10.2	2.0	60	7.24

Table 6. Shear test results of neat and CNT doped T800/M21 specimens.

	w (mm)	t (mm)	G12 (GPa)
Neat T800/M21	20.4	2.0	5.01
CNT doped T800/M21	20.4	2.0	5.6

CNT treated T800/M21 specimen is %43 higher than untreated samples [34] .

3.3. Electrical Conductivity

The results of test performed in room temperature are presented in Table 7. Addition of CNT increased the electrical conductivity of the carbon/epoxy composite significantly in fibre and thickness direction.

Table 7. Electrical conductivity of neat and CNT doped T800/M21 specimens at room temperature.

Samples	Direction	Length (mm)	Width (mm)	Thickness (mm)	Conductivity (S/m)
Neat T800/M21	Longitudinal	50.3	25.4	4.3	16278.3
	Transverse	50.1	21.5	4.2	18.2
	Thickness	50.2	22.1	4.2	1.3
CNT doped T800/M21	Longitudinal	50.3	25.2	4.4	25941.0
	Transverse	50.1	24.5	4.5	19.9
	Thickness	50.2	23.5	4.4	2.5

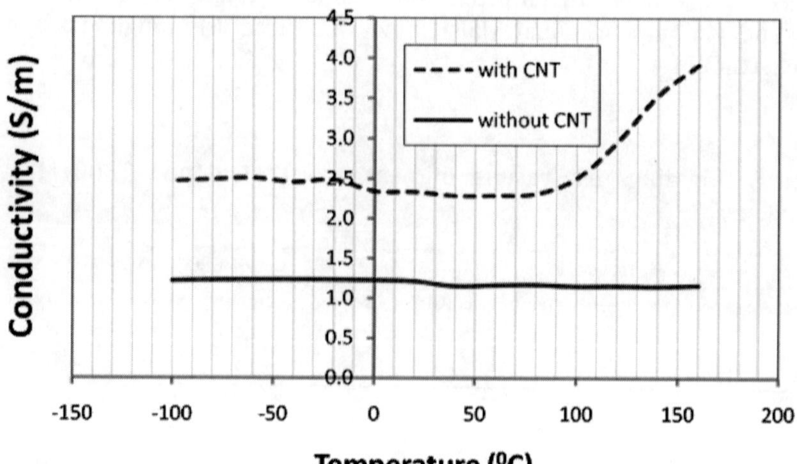

Figure 13. Temperature effect on electrical conductivity of neat and CNT doped T800/M21 specimens in thickness direction.

This increment is 59% in longitudinal direction and 92% in thickness direction. However, the increase in transverse direction is only 10% which is less than the other directions.

In Figure 13, the effect of temperature on electrical conductivity in thickness direction is given for the temperature range of −100°C and 160°C at 1 kHz frequency. According to results, the effect of temperature on electrical conductivity of neat specimens is negligible for the temperature range considered. On the other hand, the electrical conductivity increases after 100°C in CNT doped specimens significantly. Increased electrical conductivity through the thickness at high temperatures is a desirable property for lightning strike protection of aircraft structures because a sudden rise of temperature is observed at the region where the lightning hits the surface when the electrical conductivity of the material is lower. Increased electrical conductivity at elevated temperatures will further reduce the structural damage.

4. CONCLUSIONS

The thermal, mechanical and electrical behaviours of CNT doped unidirectional T800/ M21 carbon fibre/epoxy prepreg composite material are obtained by experimental studies. CNT-doped thermoplastic-based dry powder was deposited uniformly on top of prepreg material for the treatment and it increases material mass by only 1%.

In order to investigate thermal behaviour of the composite materials, differential scanning calorimetry and thermogravimetric analyses are performed to the specimens. Later, detailed thermal analysis is conducted for with/without carbon nanotube reinforced material to obtain the thermal conductivity, specific heat and thermal expansion coefficient of the material. According to the results, it is seen that the CNT reinforcement has both positive and negative effects on the thermal properties. 1) CNT addition reduces the thermal conductivity in three directions slightly whereas the glass transition and decomposition temperature increase by 5°C, which extends the upper service temperature limit of the material. The reduction of thermal conductivity is negligible in longitudinal and thickness directions whereas 12% in transverse direction. With the supporting evidence in the literature, it is concluded that the presence of CNTs in the epoxy matrix causes discontinuities in polymer leading to reduction of mean free oscillation of polymer molecules and electron scattering and in turn decreasing the thermal conductivity. On the other hand, presence of CNTs in epoxy matrix prevents movement of polymer molecules and diffusion. Thus, it leads to an increase in the glass transition and decomposition temperatures. 2) The specific heat capacity of CNT doped material decreases with the temperature by 7% compared to that of neat material. Because, CNTs situated in the epoxy hinder the movement of polymer molecules and cause a decrease in total kinetic (thermal) energy of the system. 3) CNT addition causes a decrease in expansion of specimen compared to the neat material in longitudinal direction which is considered as a positive effect for thermal stability of the material. On the other hand, the coefficient of thermal expansion of the doped specimen is significantly higher than the neat specimen in transverse direction. The effect is less significant at elevated temperatures.

The effect of carbon nanotube reinforcement on mechanical behaviour is studied by tensile, bending and shear tests. The results show that the addition of multi-walled carbon nanotubes to the CFRP improves all mechanical properties because it prevents the formation and propagation of cracks that affect load distribution significantly. CNTs show the greatest effect in the failure strain in longitudinal direction and in the tensile strength in transverse direction. Addition of CNT in carbon/epoxy composite system increases the tensile strength by 7%, the elastic modulus by 4%, and the failure strain by 16% in the longitudinal direction. In the transverse direction, the tensile strength increases by 21%, the elastic modulus increases by 21% and the failure strain increases by 9%. The addition of CNTs also increases the bending modulus by 11% for both fibre and transverse directions.

It is found that the addition of carbon nanotube increases the electrical conductivity in all directions. In CNT doped specimens, electrical conductivity of the carbon/epoxy composite increases significantly in longitudinal direction by 59% and thickness direction 92% whereas less increase of 10% is observed in transverse direction. The electrical conductivity increases after 100°C for the doped material in thickness direction, which is desirable property for aircraft structures subject to lightning strike in order to reduce the corresponding damage area further. According to the studies in SARISTU, micro- and nanoparticle addition has been successfully used to improve the potential discharge time and to allow otherwise isolating components to perform low-level electrical tasks. CNTs with high aspect ratio particles reduced the required filler content for electrical percolation. Therefore it is beneficial to include CNT in CFRP to reduce the number and size of electrical structural network components. Requirements for lightning strike direct effects are difficult to achieve with CNT doped prepreg to provide a stand-alone solution for electrical conductivity. However, they can help to reduce material damage.

ACKNOWLEDGEMENTS

SARISTU project has received funding from the European Union's Seventh Framework Programme for research; technological development and demonstration under grant agreement no 284562.

REFERENCES

1. Yu, H., Lu, C., Xi, T., Luo, L., Ning, J. and Xiang, C. (2005) Thermal Decomposition of the Carbon Nanotube/SiO2 Precursor Powders. Journal of Thermal Analysis and Calorimetry, 82, 97-101.

2. Chang, C.-W., Tseng, J.-M., Horng, J.-J. and Shu, C.-M. (2008) Thermal Decomposition of Carbon Nanotube/Al2O3 Powders by DSC Testing. Composites Science and Technology, 68, 2954-2959.

3. MCguire, N.K. (2002) Negative-Coefficient Materials Can Point the Way to Positive Value in the Right Matrixes. http://pubs.acs.org/subscribe/archive/tcaw/11/i11/pdf/1102 mcguire.pdf

4. HexPly® M18/1. http://www.hexcel.com/Resources /DataSheets/Prepreg-Data-Sheets/M18_1_eu.pdf

5. Torayca T800H Data Sheet. http://www.toraycfa.com /pdfs/T800HDataSheet.pdf

6. Rios, P., Kenig, S., Cohen, R. and Shechter, A. (2013) The Effect of Carbon Nanotubes on the Thermal Expansion Isotropy of Injection Molded Carbon Fiber Reinforced Thermoplastics. Polymer Composites, 34, 1367-1374.

7. Sathyanarayana, S. and Hübner, C. (2013) Structural Nanocomposites. Springer, Berlin.

8. Flores, S. and Gayoso, J. (2016) Enhancement of Primary Structure Robustness by Improved Damage Tolerance. In: Wolcken, P.C. and Papadopoulos, M., Eds., Smart Intelligent Aircraft Structures (SARISTU), Springer, Cham, 763-775.

9. Choudhary, V. and Gupta, A. (2011) Carbon Nanotubes—Polymer Nanocomposites. InTech, Rijeka.

10. Mittal, V. (2010) Polymer Nanotube Nano-composites: Synthesis, Properties and Applications. Wiley-Scrivener, New Jersey.

11. Pistor, V., Ornaghi, F.G., Ornaghi Jr., H.L. and Zattera, A.J. (2012) Degradation Kinetic of Epoxy Nanocomposites Containing Different Percentage of Epoxycyclohexyl-POSS. Polymer Composites, 33, 1224-1232.

12. Levchik, S.V. and Weil, E.D. (2004) Thermal Decomposition, Combustion and Flame-Retardancy of Epoxy Resins—A Review of the Recent Literature. Polymer International, 53, 1901-1929.

13. Loos, M.R., Coelho, L.A.F., PezzinI, S.H. and Amico, S.C. (2008) The Effect of Acetone Addition on the Properties of Epoxy. Polímeros, 18, 76-80.

14. Pistor, V., Soares, B.G. and Mauler, R.S. (2012) Kinetic Degradation of Hybrid Polymer Nanocomposites.

15. Mouritz, A. and Gibson, A. (2006) Fire Properties of Polymer Composite Materials. Springer, Dordrecht.

16. Tong, X.C. (2011) Advanced Materials for Thermal Management of Electronic Packaging. Springer Series in Advanced Microelectronics 30. Springer, New York.

17. Siddiqui, N.A., Khan, S.U., Ma, P.C., Li, C.Y. and Kim, J.-K. (2011) Manufacturing and Characterization of Carbon Fibre/Epoxy Composite Prepregs Containing Carbon Nanotubes. Composites Part A: Applied Science and Manufacturing, 42, 1412-1420.

18. Kim, S., Kim, J.T., Kim, H., Rhee, K. and Kathi, J. (2012) Thermal and Mechanical Properties of Epoxy/Carbon Fiber Composites Reinforced with Multi-Walled Carbon Nanotubes. Journal of Macromo-lecular Science, Part B: Physics, 51, 358-367.

19. Chang, M.S. (2010) An Investigation on the Dynamic Behavior and Thermal Properties of MWCNTs/FRP Laminate Composites. Journal of Reinforced Plastics and Composites, 29, 3593-3599.

20. Lin, Y., Gigliotti, M., Lafarie-Frenot, M.C. and Bai, J. (2015) Effect of Carbon Nanotubes on the Thermoelectric Properties of CFRP Laminate for Aircraft Applications. Journal of Reinforced Plastics and Composites, 34, 173-184.

21. Lonjon, A., Demont, P., Dantras, E. and Lacabanne, C. (2012) Electrical Conductivity Improvement of Aeronautical Carbon Fiber Reinforced Polyepoxy Composites by Insertion of Carbon Nanotubes. Journal of Non-Crystalline Solids, 358, 1859-1862.

22. Moisala, A., Li, Q., Kinloch, I. and Windle, A. (2006) Thermal and Electrical Conductivity of Single- and Multi-Walled Carbon Nanotube-Epoxy Composites. Composites Science and Technology, 66, 1285-1288.

23. Gojny, F.H., Wichmann, M.H., Fiedler, B., Kinloch, I.A., Bauhofer, W., Windle, A.H. and Schulte, K. (2006) Evaluation and Identification of Electrical and Thermal Conduction Mechanisms in Carbon Nanotube/Epoxy Composites. Polymer, 47, 2036-2045.

24. Kostopoulos, V., Baltopoulos, A., Karapappas, P., Vavouliotis, A. and Paipetis, A. (2010) Impact and After-Impact Properties of Carbon Fibre Reinforced Composites Enhanced with Multi-Wall Carbon Nanotubes. Composites Science and Technology, 70, 553-563.

25. Ramirez, V., Hogg, P. and Sampson, W. (2015) The Influence of the Nonwoven Veil Architectures on Interlaminar Fracture Toughness of Interleaved Composites. Composites Science and Technology, 110, 103-110.

26. Gojnya, F.H., Wichmanna, M.H., Fiedler, B., Bauhofer, W. and Schulte, K. (2005) Influence of Nano-Modification on the Mechanical and Electrical Properties of Conventional Fibre-Reinforced Composites. Composites Part A: Applied Science and Manufacturing, 36, 1525-1535.

27. Allaouia, A., Baia, S., Cheng, H. and Bai, J. (2002) Mechanical and Electrical Properties of a MWNT/Epoxy Composite. Composites Science and Technology, 62, 1993-1998.

28. Ayatollahi, M.R., Shokrieh, M.M., Shadlou, S., Kefayati, A.R. and Chitsazzadeh, M. (2011) Mechanical and Electrical Properties of Epoxy/Multi-Walled Carbon Nanotube/Nanoclay Nanocomposites. Iranian Polymer Journal, 10, 835-843.

29. Alva, A. and Raja, S. (2011) Dynamic Characteristics of Epoxy Hybrid Nanocomposites. Journal of Reinforced Plastics and Composites, 30, 1857-1867.

30. Bal, S. (2010) Experimental Study of Mechanical and Electrical Properties of Carbon Nanofiber/Epoxy Composites. Materials and Design, 31, 2406-2413.

31. Feller, J., Linossier, I. and Grohens, Y. (2002) Conductive Polymer Composites: Comparative Study of Poly(ester)-Short Carbon Fibres and Poly(epoxy)-Short Carbon Fibres Mechanical and Electrical Properties. Materials Letters, 71, 57-64.

32. Ogasawara, T., Hirano, Y. and Yoshimura, A. (2010) Coupled Thermal-Electrical Analysis for Carbon Fiber/Epoxy Composites Exposed to Simulated Lightning Current. Composites Part A: Applied Science and Manufacturing, 41, 973-981.

33. Kinloch, A.J.R., Mohammed, D., Taylor, A.C., Eger, C., Sprenger, S. and Egan, D. (2006) The Effect of Silica Nano Particles and Rubber Particles on the Toughness of Multiphase Thermosetting Epoxy Polymers. Journal of Materials Science, 4, 1293-1307.

34. Florez, S., Gaztelumendi, I. and Gayosa, J. (2016) Smart Intelligent Aircraft Structures (SARISTU). Springer, London

2

ALIGNED CARBON NANOTUBE–EPOXY COMPOSITES: THE EFFECT OF NANOTUBE ORGANIZATION ON STRENGTH, STIFFNESS, AND TOUGHNESS

Anastasiia Mikhalchan[1], Thurid Gspann[1], and Alan Windle[1]

Macromolecular Materials Laboratory, Department of Materials Science and Metallurgy, University of Cambridge, 27 Charles Babbage Road, Cambridge CB3 0FS, UK

ABSTRACT

A protocol has been developed for the production of epoxy-based composites containing high-volume fractions of aligned carbon nanotubes. The nanotubes were fabricated as continuous fibres or aligned mats directly from the CVD reactor, in which they were synthesized. The block composites with highly aligned and tightly packed nanotube assemblies were prepared via epoxy resin infiltration, and their volume fraction, distribution, and internal porosity being analysed prior to mechanical testing. The samples were tested in both axial tension and three-point bending. The results show that the strength and stiffness enhancements were close to pro rata with the volume fraction of the carbon nanotubes added. The failure modes were distinctly different from those characteristic of the conventional aligned carbon fibre composites. The fracture surface showed considerable evidence of pull-out of bundles of (~50) nanotubes, but the pull-out appeared to involve the resin matrix which drew out along with the bundles. Subsidiary cracks were bridged by nanotube bundles giving structures reminiscent of crazes in glassy polymers, what constitutes the distinct toughness mechanism and higher resistance to the transverse cracks propagation.

INTRODUCTION

Materials and methods

Materials

Carbon nanotubes were synthesized in a CVD tube reactor under hydrogen atmosphere using toluene as carbon precursor, ferrocene as catalyst, and thiophene to provide the sulfur 'promoter'. The nanotubes were collected as macro-assemblies—either in the form of fibres or mats. For fibres, the CNT aerogel was spun out of the gas phase at rates of 10–30 m/min and condensed on-line by atomized acetone, the stage at which the open network of nanotube bundles merges to form a more continuous yarn-like fibre with a uniform cross section (which we refer to as CNT fibre or CNTF). The yarn-like nature of the fibre arises from its composition of 10^4 to 10^5 CNT bundles per cross section, each of which is relatively 'slippery' with respect to its neighbours [15]. During the condensation but prior to epoxy infiltration, about 1000 individual fibres, each with an average tensile strength of 1.16 GPa [16], linear density of 0.03 tex, and 8 μm diameter were aligned and wound up together to form 'macro-fibres' (or 'unidirectional tows') of ~200 μm diameter. CNT mats (CNTM) were synthesized under the same conditions, but wound up continuously without acetone condensation.

Composites were made using a hot curing epoxy matrix based on bisphenol A epoxy resin (Araldite LY 1564), anhydride hardener (Aradur 917), and amine accelerator (960-1). The epoxy system is recommended for industrial filament winding, pultrusion, wet lay-up, or Resin Transfer Moulding (RTM) manufacturing processes to produce composites with enhanced mechanical properties.[1] It combines a low mix viscosity (100–200 mPa·s at 40 °C), for efficient penetration into our nanotube material, with a pot life at impregnation temperature of several hours. In line with normal composite practice, 1.5 % wt of a commercially available surfactant (BYK-Chemie GmbH) was added to the matrix as an 'air release' agent.

Nanocomposites processing

The formulated epoxy system was mixed thoroughly to ensure homogeneity at a mix ratio 100 × 98 × 3 (resin × hardener × accelerator, % by weight) as recommended by the supplier, then degassed in a vacuum oven for 30–60 min at 40 °C before impregnation. Pre-determined amounts of carbon nanotubes, whether in the form of fibres or aligned mats were then, after orientation with respect to the long axis of the custom-built mould, infiltrated with matrix material under pressure at 40 °C, and degassed before curing. All specimens, including the control sample of pure matrix, were cured at the same cycle: 3 h at 100 °C, then 2 h at 120 °C in a vacuum oven, followed by oven cooling to room temperature. Double-sided silicon release paper was used as anti-adhesive between sample and mould. All the samples were carefully polished prior to further investigations.

Characterization methods

TGA thermograms were recorded (TA Instruments Q500) by heating the specimens from room temperature to 700 °C in a dynamic oxidative atmosphere (air) with a flow rate of 60 mL/min. These measurements were used to verify weight fractions of the nanotube material in the composites using the protocol described in [12]. The apparent volumetric density of each sample was calculated from its dimensions taken by a digital micrometer and its mass measured using high-precision Sartorius microbalance. X-ray micro-tomography (Bruker SkyScan 1172) was used for microstructural characterization and three-dimensional (3-D) visualization (details in supplementary materials, S1). SEM images were obtained with the FEI Nova NanoSEM 450. All specimens were coated with platinum for 1 min to minimise charging. Cross sections of the CNT fibres and composites were cut using a Focused Ion Beam (FIB) using the Helios FIB/FEG-SEM workstation. The 2D FFT analysis of the SEM images (grayscale 8-bit, cropped to 1024 × 1024 pixels) was done with the ImageJ software (developed by the National Institutes of Health) and an oval-profile plug-in (authored by Bill O'Connell) available from the public domain http://rsb.info.nih.gov/ij [17]. Polarized Raman spectra were obtained with a Bruker Senterra Raman microscope using 785-nm excitation (supplementary materials, S2). The maximum intensity ratios of IG_0/IG_{90} (I_{VV}/I_{HH}, with the sample being aligned with the incident beam in V direction) were each calculated by averaging a map of ten positions.

Three-point bend tests were performed with a Hounsfield test machine model 500 L with the 250-N load cell, a span-to-depth ratio of 40:1, and a speed rate of 1 mm/min. Each specimen was deflected until rupture occurred. The load–deflection curves, after removal of the initial 'toe region', were used to determine the flexural strength, strain, and modulus of elasticity according to the ASTM D 7264 standard.

Tensile tests according to the ASTM D7205/D7205M-06 were obtained using an Instron test machine model 859i with pneumatic grips, a 100-kN load cell, a test speed of 10 mm/min, and an initial gauge length of 20 mm. To prevent slipping while avoiding stress concentrations at the grips, both ends of each specimen were glued in between of two Hexlite panels (glass fibre-reinforced honeycomb, Hexcel Ltd.) using an epoxy bi-componential rapid adhesive glue, and held overnight at room temperature under pressure for optimum bonding.

RESULTS

Carbon nanotube assemblies

Carbon nanotube fibres and mats were used to compare their reinforcement efficiency for epoxy matrix. Both mats and fibres were composed of mostly double-wall nanotubes collapsed and formed into bundles of 40–50 nm diameter and length of the order of 100 μm. According to TGA, the residual Fe-catalyst varied from 1 to 3.7 % wt (see supplementary materials, S3). SEM analysis confirmed a high degree of preferred axial orientation of the nanotubes, their overall uniformity, and the absence of any large impurity clusters (Fig. 1).

The longitudinal and cross sections of the fibre (Fig. 1a, b) show that, despite condensation, the structure contains a considerable amount of voids between the nanotube bundles. They are highly elongated and interconnected with the result that the resin penetrates the condensed fibres and fills the voids [12, 13]. The interconnected voids are also confirmed by the specific surface area of the fibre which can be up to 200 m^2/g depending on the synthesis conditions [4,13, 14]. The oriented CNT mat comprises a far less dense network of laterally entangled aligned nanotube bundles (Fig. 1c). We should emphasize that all CNT mats and fibres were incorporated into the composites as-synthesized and without any further modification by mechanical or chemical post-treatments.

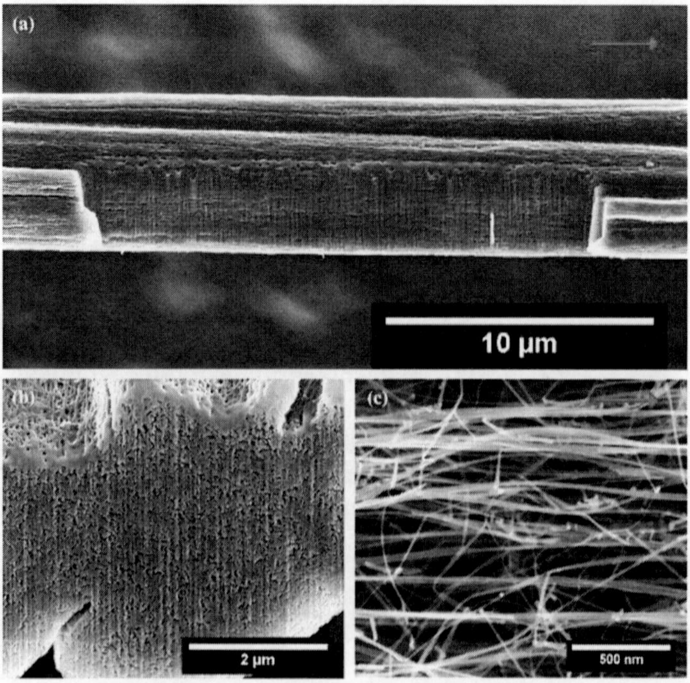

Figure 1. SEM images of CNT assemblies used as reinforcing components of the composites: an individual condensed CNT fibre with its **a** longitudinal- and **b** cross-sectional cut by Focused Ion Beam (FIB); and**c** uncondensed CNT mat. *Red arrows* indicate winding direction and bundle orientation. The *'melt-like' top surface edges* and the *vertical lines* on the FIB sections are ion beam cutting artifacts

The orientation of CNT bundles within their assemblies can be quite clearly seen from the exemplary SEM images (Fig. 2a, d). For the uncondensed CNT mats, the CNT bundles form a continuous network; however, some types of bundles misalignment are visible, including their local curvature, twisting, loops, or changing direction perpendicularly to join a neighbouring bundle. For the CNT fibres, bundles and inter-bundle pores are mostly oriented in the fibre's axis (synthesis and winding direction), which is visible on both top and longitudinal section surfaces at high magnification. It is obvious that liquid densification causes merging of CNT bundles, enhancing their packing efficiency in the fibre and resulting in local curvature which is less distinct than in the uncondensed CNT mats. The less local curvature (so-called "waviness") was also observed for the vertically aligned CNT arrays upon their mechanical densification [18].

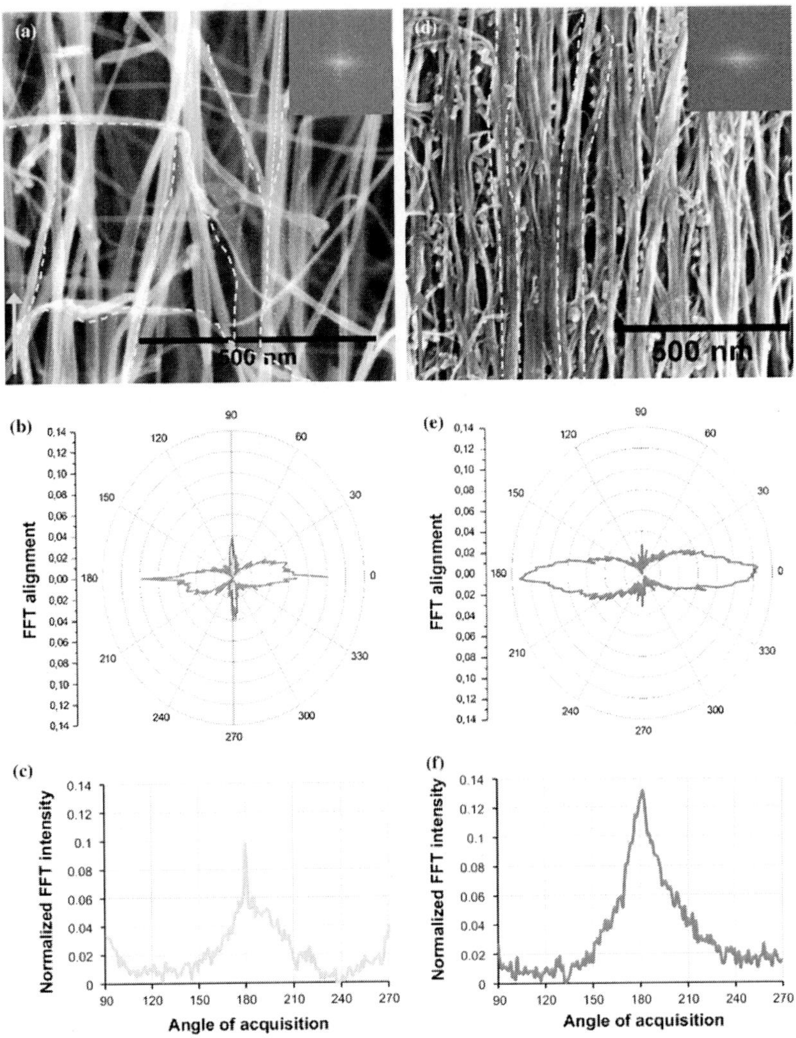

Figure 2. Analysis of the CNT bundles alignment within their assemblies: representative SEM images of **a** an uncondensed CNT mat, **d** top surface of a condensed CNT fibre at high magnification. *Scale bars* are 500 nm. *Yellow dash lines* highlight some visible misalignments and local curvature of CNT bundles. *Green and red arrows* show the synthesis and winding direction (fibre axis). Corresponding *insets* to **a**and **d** are the 2D FFT patterns used for radial integration of pixel intensities (**b**, **e**). **c**, **f** represent the azimuthal scans summed over a range of radii used for the calculation of the first-order spherical harmonic coefficient

The differences in alignment apparent in the micrographs are quantified using the two-dimensional Fast Fourier Transform (2D FFT) analysis of the representative SEM images. For this purpose, the images were transformed into binary version composed exclusively of black and white pixels, thus reducing the noise effects and spurious information. The FFT patterns (Fig. 2 insets a, d). were then radially integrated to obtain FFT intensity profile as a function of angle of acquisition, I(ϕ) (Fig. 2b, c, e, f). The base line of this plot was taken as the minimum recorded value of intensity.

Values of the first even spherical harmonic coefficient were calculated for the cases of CNT fibre and CNT mat from the plots of $I(\phi)$ (Fig. 2c, f). As the harmonic coefficient refers in this case to orientation observed in the plane of the micrograph, whereas its designation as P2 implies cylindrical symmetry the fibre axis, something which certainly cannot be assumed in the case of the mat, we refer to it as the Herman's orientation function, *H*, defined, as for P2, as:

$$H = \frac{3 \cdot <(\phi) \cdot \cos^2 \Phi > -1}{2}$$

where the diamond brackets denote the average value of the product over equal angular intervals normalized to the average intensity.

It is interesting to note that the fraction of bundles in the mat sample which are not well aligned tends to be at around 90° to the main alignment axis, indicating that the bundles had been subject to some spreading action possibly during wind up. This population produces the subsidiary peaks, at ±90° from the main peak (Fig. 2c), which will tend to reduce the value of *H*. The values for *H* obtained are: mat 0.5 and fibre 0.63.

The actual value of H is dependent on the choice of base line. For example, if most of the CNTs had random orientation with only a small sub-set providing the particular distribution, then the values of P2 would have been less. We know that there is no such background from the micrographs, so taking the base line as the minimum value data point, but it would be good to have further confirmation that this is correct. To this end, we carried out polarized Raman measurements of orientation measuring the relative intensities of the G peak, with polarizer and parallel analyser aligned first with the fibre axis and then normal to it, known as VV and HH configuration. The data are shown in Fig. 3. It has been demonstrated [19] that when the direction of polarization is perpendicular to a single-wall nanotube, the intensity of the G peak is too small to measure. The result of this is that measurements of the G peak intensity of our samples in the 0° and 90° positions provide the ratio expected for the intensity at these two angles shown in the image analysis plots (Fig. 2c, f) and, thus, a check on the baseline choice for the FFT analysis of the micrographs.

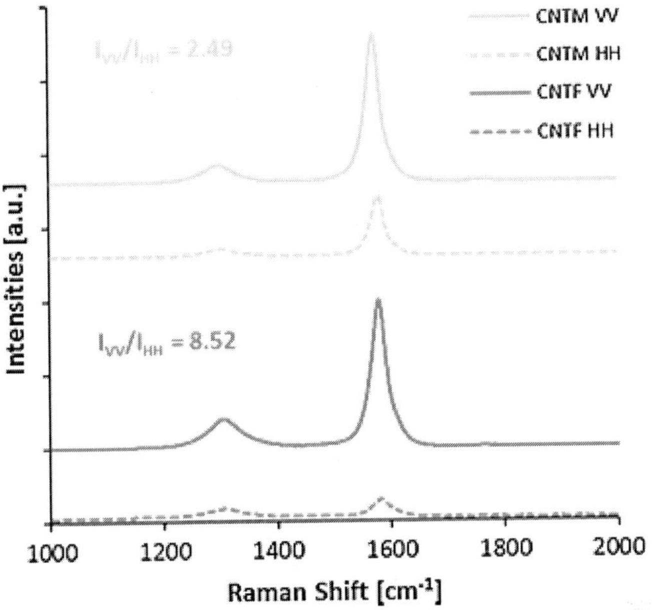

Figure 3. Characterization of CNT alignment in mats and fibres by polarized Raman spectroscopy. Raman spectra in VV and HH configuration, with the spinning direction/CNT bundle orientation being parallel to the V configuration of the incident beam, of CNT fibres (*red*) and **b** CNT mats (*green*). The vertical scaling is same for each curve, the curves have been displaced for clarity. Each of the *plots* is the average of ten measurements made over different parts of the sample

The ratios are as follows:

I_0/I_{90} mat sample	2.85 (±0.5)/1—image analysis (Fig. 2).
	2.49 (±0.1)/1—Raman (Fig. 3).
I_0/I_{90} fibre sample	7.8(±1.2)/1—image analysis (Fig. 2).
	8.52 (±0.2)/1—Raman (Fig. 3).

Given the generous error bands on the estimation of the I_0 and I_{90} values from the FFT analysis plots, we have not made any baseline adjustments which would have an effect on H.

For post-spinning condensed mats, I_{VV}/I_{HH} intensity ratio increased to ~3.5 which corroborates our observations of the effect of condensation on mechanical properties (supplementary materials, S4), where condensation enhanced lateral contacts between the CNT bundles as well as causes some local alignment.

Nanocomposite structure

Figure 4a shows a typical nanocomposite sample with dimensions of 40 × 2 × 0.2–0.8 mm. 3D (Fig. 4b, c), and 2D 'slice' images (Fig. 4d) reconstructed from X-ray micro-tomographic data show the uniform, highly anisotropic structure of a CNTF-composite with near perfectly aligned CNT fibres. The images confirm the absence of voids at its resolution limit of just below 1 μm, although the isolated small bright spots may indicate some sub-micrometer voids.

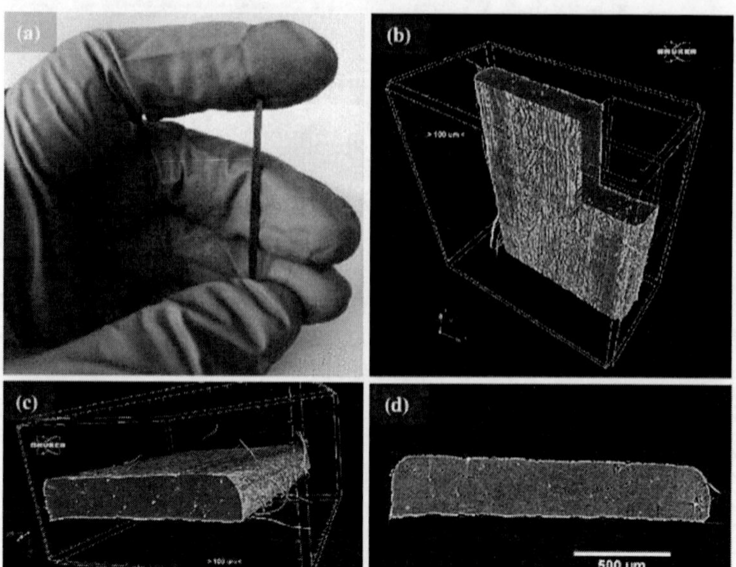

Figure 4. **a** CNTF-nanocomposite specimen, **b** and **c** X-ray tomography 3-D reconstruction, and **d** 2-D slice image of a representative CNTF-epoxy nanocomposite. Composites made from mats rather than CNT fibres show similar tomographic results

The weight fraction of the carbon nanotubes for each sample was measured by TGA according to the procedure described previously [12], and showed a high level of agreement with the known weight of nanotubes incorporated in each sample (see supplementary materials, S5), while the density of the nanotube bundles, in which the nanotubes are well aligned and packed, is calculated in supplementary materials, S6. We use TGA-based weight fractions for the data below.

Figure 5 shows that the calculated density values, while initially following the theoretical line, deviate for CNT weight fractions in excess of 40 %, as the measured density increasingly falls below the predicted line, a trend which we ascribe to an increasing percentage of voids in the composite. Complete penetration of the densely packed nanotube bundle network with resin appears to become more difficult as the CNT fibres are packed to greater densities in the mould. The data of Fig. 5 can thus be expressed as void content as in Fig. 6a below (for evaluation of the volume fraction and void content in composites refer to the supplementary materials, S7).

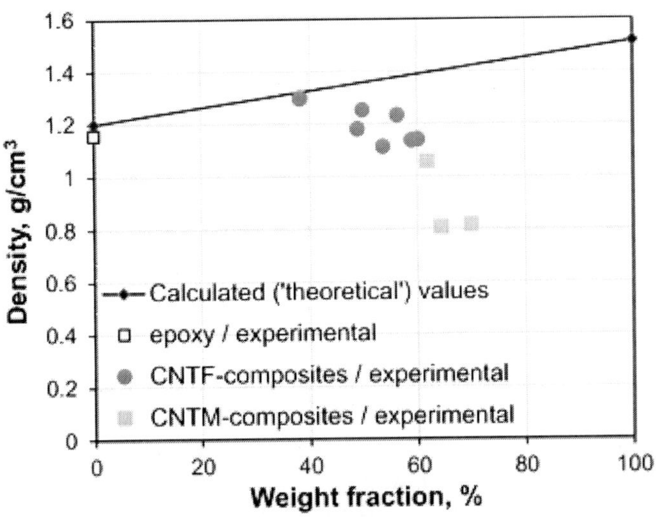

Figure 5. Comparison between theoretical and actual densities of the samples depending on weight fractions. The zero weight fraction is attributed to pure epoxy resin, while 100 % corresponds to the density of pure CNTs. The *black solid line* represents the 'theoretical' density calculated from TGA-verified weight fractions and densities of the pure components for the 'ideal' composite without voids, with the density of the nanotube bundles being calculated as in (supplementary materials, S6). *Coloured points*show the actual density data determined from mass and volume measurements

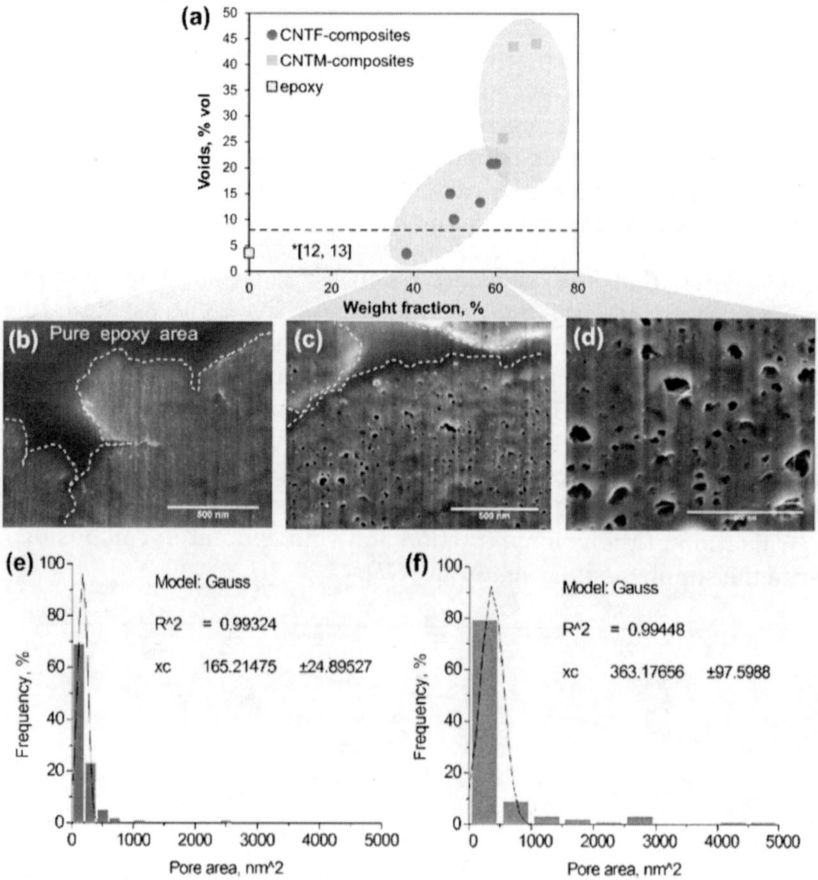

Figure 6. **a***Graph* shows the general trend of void content increase (in % vol) with increasing CNT content (in % wt) in nanocomposites. The *horizontal dashed black line* (at 8 % vol) represents a typical value of void content in carbon fibre-reinforced composites for an out-of-autoclave cure [21, 25] or cure in a vacuum bag with zero autoclave overpressure [26]. The *yellow oval area* is based on our previous results on low CNT-content composites [12, 13]. FIB-SEM cross sections of the nanocomposites filled with **b** and **c** CNT fibres and **d** CNT mats also show increasing void content at higher CNT loadings.*Yellow dashed lines* on images **b** and **c** delineate the infiltrated CNT fibres from inter-fibre regions of pure epoxy. **e, f** show histograms of the pore-size distribution determined from **c** and **d** images together with their Gaussian fits

Intrinsic porosity and infiltration

The most common mechanisms for the formation of micro-voids or even macro-defects in the matrix are moisture absorption by resin components or prepregs on storing [20], release of volatile gases and water vapors in the curing process [21, 22], air-trapping and lay-up imperfections [20, 22, 23], or simply handling issues [24]. Voids of different shapes and sizes tend to occur primarily at the interface between the plies of prepregs, or between an individual reinforcement element and surrounding polymer matrix.

For CNT-composites, the void content becomes substantial on approaching CNT contents of ~70 % by weight (Fig. 6a).

SEM analysis of the cross section of a specimen with low filler content, thus cured in excess of epoxy (as reference only, not tested in this work), indicates easy-detectable regions of pure epoxy matrix and very good fibre impregnation (Fig. 6b). This observation comes in full agreement with the previous studies from this group, where epoxy completely filled the gaps between CNT bundles in the fibres in composites with 17–26 [12] and 5–33 % [13] weight fractions.

Interestingly, the cross sections of composites with higher CNT loadings show a higher void content located primarily inside the fibres as nano-scale pores, but not at the fibre–fibre or fibre–epoxy interface (Fig. 6c). Gaussian-fit histograms of the pore-size distribution indicate that for CNTF-nanocomposites, 70 % of all pores are smaller than 250 nm^2 and have an average area of 165 nm^2 (Fig. 6e). This value is almost a half as that of the CNTM-composites, where an average pore area is 363 nm^2 (Fig. 6f).

These data suggest that infiltration of the CNT fibres with resin is a key aspect of the nanocomposite material—which obviously does not occur for carbon fibre composites—and will have implications for the future optimization of the processing protocols.

Figure 6c also shows that at high-volume content of CNTs in the composite (we have used the volume fractions re-calculated from the experimentally verified weight fractions with the estimated CNT bundles density coupled with the fact that epoxy does not penetrate the bundles; the measured void fraction is also considered, supplementary materials, S5–S7), the fibres appear to be packed very densely with each other almost without visible layers of epoxy between them; however, this is not surprising. Considering a 1-m length CNT fibre with 0.028 tex and diameter of about 7 µm, its density simply calculated as a mass-to-volume ratio is only 0.73 g/cm^3. If considered density of the CNT bundles of about 1.66 g/cm^3, thus the void content within the CNT fibre itself appears to be approximately 50 % (supplementary materials, S8), which is very close to what we observed in our previous work [14]. For practice, it means, that the volume of resin occluded with the fibre is equal to volume occupied by CNT bundles within the fibre. Hence, 30–40 % volume loading of CNTs is relative and equivalent to about 60–80 % of the composite's total volume being occupied by the CNT fibres *plus* infiltrated resin. It confirms high actual packing density, thus we observe fibres packed very densely almost without visible regions of pure epoxy in between them. In contrast, our previous work [12, 13] and the sample repeated as a reference for the same weight fraction (Fig. 6b) have shown that for the composites containing 26–33 % wt carbon nanotubes, CNT fibres were not packed so tightly, with large regions of pure epoxy matrix between them (refer also to Fig. 3 in [12], and Figs. 1 and 2 in [13]).

Behavior in tension

Nanocomposites with similar volume fractions of CNT mats and fibres were tested to compare the tensile properties of the two morphologies with pure epoxy matrix. Figure 7 shows representative stress–strain curves, and Table 1 summarizes data of tensile strength, stiffness, and elongation-to-break. For better comparison with the fibrous reinforcing elements, specific values of strength and stiffness of the composites were also calculated in N/tex which is equivalent to the units of GPa/density (density in g/cm^3). The density in g/cm^3 is often expressed as Specific Gravity (SG) which is the density relative to that of water, while linear density is mass per unit length as tex (g/km).

The improvement in strength and stiffness for the composite based on condensed CNT fibres is notable, being at least pro rata (linearly proportional to volume fraction) on the basis of the strength of individual fibres. For 35 % volume loading, the specific tensile strength of the epoxy is enhanced a factor of ~5 and the specific modulus by a factor of ~7.9. The influence of uncondensed CNT mats as a reinforcing agent seems initially much less impressive, with enhancement factors of only ~2.7 and ~4, respectively. We ascribe the larger enhancement in the case of CNTF to the fact that the condensation considerably improves the degree of mutual alignment and closer contacts of the CNT bundles (supplementary materials, S4).

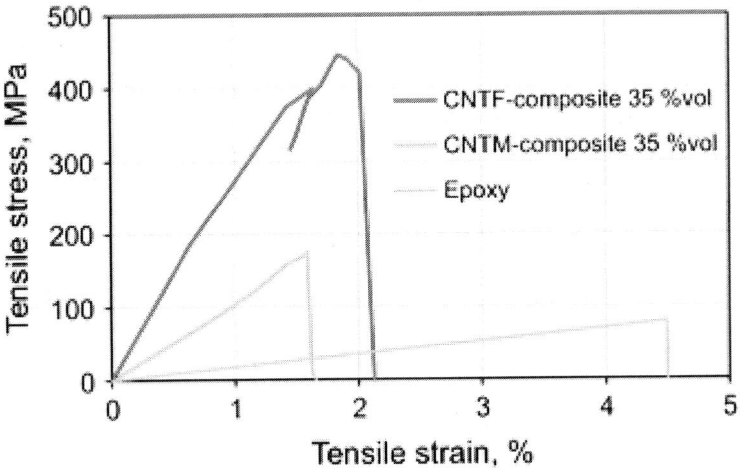

Figure 7. Representative tensile stress–strain curves for pure epoxy matrix and nanocomposites

Table 1. xial tensile properties of the pure epoxy, CNT fibres, mats, and their composites

Sample	Volume fraction (%)[a]	Strength		Stiffness		Elongation (%)
		MPa	GPa/SG (N/tex)	GPa	GPa/SG (N/tex)	
Pure epoxy	0	81.5 ± 6.5	0.076 ± 0.006	3.1 ± 0.1	2.9 ± 0.1	4.5 ± 0.5
CNTF-composite	35	435 ± 10	0.390 ± 0.009	25.5 ± 1	22.9 ± 0.9	2.6 ± 0.2
CNTM-composite	35	170 ± 9.5	0.206 ± 0.011	9.5 ± 1	11.5 ± 1.2	1.6 + 0.2
CNT fibre 0.028 tex	100	–	1.16 ± 0.19	–	55.7 ± 11.8	3.6 ± 0.8
CNT mat 0.964 tex	100	–	0.48 ± 0.12	–	7.6 ± 1.5	10 ± 0.7

[a] Volume fractions were calculated from the assumption that a composite is composed of volumes occupied by CNTs, epoxy matrix, and voids, thus including the void content added into the matrix (supplementary materials, S7)

Less perfect alignment would explain the considerably greater extension to fracture, and the lower axial modulus, in the case of the composites based on CNT mats as opposed to fibres. Recent simulations of CNT vertical arrays show that the effective modulus can be reduced by several orders of magnitude due to CNT waviness [27, 28]. As-grown less-packed CNT arrays experience a larger local curvature and large reductions in effective modulus, while mechanically densified arrays are less curved and their stiffness is dominated by the low shear modulus of CNTs [28]. In respect to composite reinforcement, there have been several papers published from the early 2000s focused mostly on composites with low volume fractions (less than 25 vol%) of short discrete nanotubes [29–31] or nanofibres [32]. It was shown that even slight nanotube curvature can significantly reduce the effective reinforcement potential when compared to straight nanotubes. The fractography of the tensile specimens after failure in Fig. 8a shows evidence of shear failure both between the fibres and within the fibres between the CNT bundles.

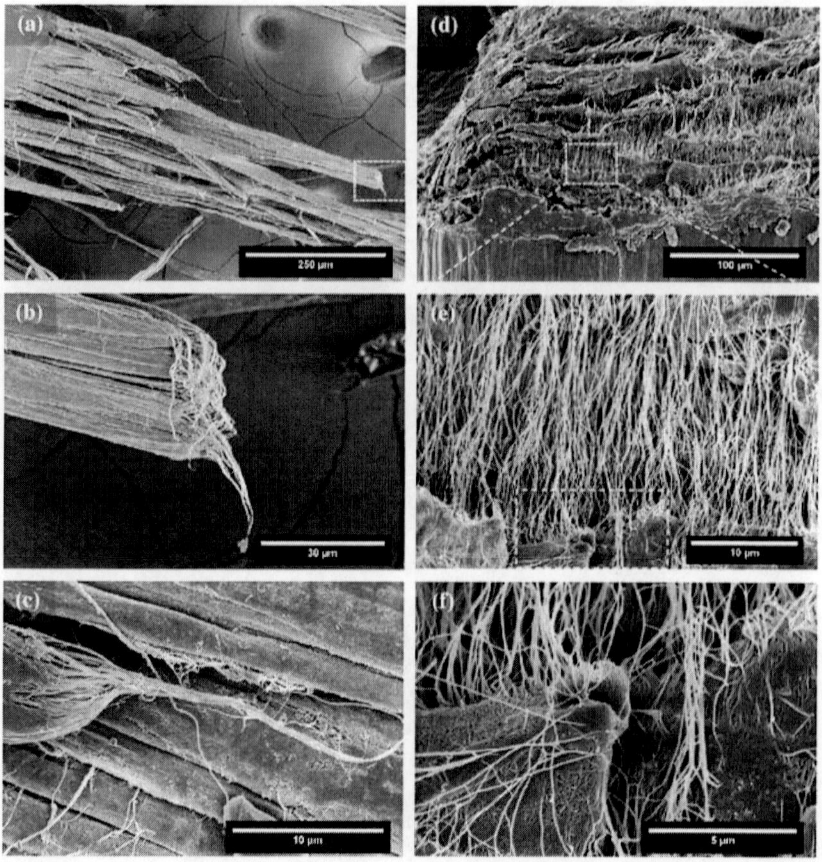

Figure 8. Fracture surfaces from tensile tests of the **a–c** CNTF- and **d–f** CNTM-composites. In the case of the composites based on condensed fibres, the sample shows evidence of failure in shear between the fibres a spacing of ~5 μm. The *inset* of **a** is magnified in **b** which shows the failure mode of several fibres, involving pull-out of the nanotube bundles. CNTM-composites show very considerable evidence of the pull-out of the nanotube bundles at the fracture surface in **d** and **e**

The fact that the tensile stress is applied to the external faces of the specimen in the grips means that the stress is only transmitted into the centre of the sample through shear, and the considerable mechanical anisotropy of the sample means that the load carried and thus extension of the individual fibres will vary with distance below the surface, with the outer fibres carrying the higher stress, at least close to the grips. This aspect has been studied and analysed in greater detail in respect of the as-spun CNT fibres by Gspann et al. [15]. It is possible that the transient load drop evident in the stress strain curve (Fig. 7) may correspond to the premature failure one of the outermost fibres, which did not trigger final failure.

For this testing regime, it appears that failure in shear between the fibres is a significant component of the total failure process and a consequence of the particular testing geometry. The fracture analysis as shown in Fig. 8b and c bears evidence of nanotube bundle pull-out, even though each fibre is, of course, impregnated with epoxy.

The fracture morphology of the sample based on CNT mats rather than condensed fibres is shown in Fig. 8d–f. It seems clear that the nanotube bundles have pulled-out from the fracture surface, although there is no evidence of them leaving holes behind in the epoxy matrix. However, in contrast to the CNTF-composite, we did not observe delamination along the specimen long axis.

Behavior in bending

The flexural properties were determined using the standard test method according to ASTM D7264 developed for continuous fibre-reinforced polymeric composites and intended to evaluate long-beam flexural behavior. According to this method, the support span-to-thickness ratio should be chosen, such that failure occurs at the outer surface of a specimen, due to only the bending moment, which may appear as a crack on the tension surface, or as the local buckling on the compression side. In the case of highly anisotropic specimens where the flexural modulus is notably reduced by shear deformation, span-to-thickness ratios higher than the standard value of 32:1 are recommended. For testing, a span-to-thickness ratio of 40:1 was used, with a specimen length of about 20 % longer than the support span distance.

The flexural strength—maximum stress at the outer surface at mid-span—and the flexural modulus of elasticity—ratio of stress to corresponding strain within the elastic limit—were calculated based on beam theory (please refer to ASTM D7264).

Representative strain–stress curves shown in Fig. 9 accompanied with data in Table 2 reveal distinctive bending behavior of the nanocomposites with the two morphologies.

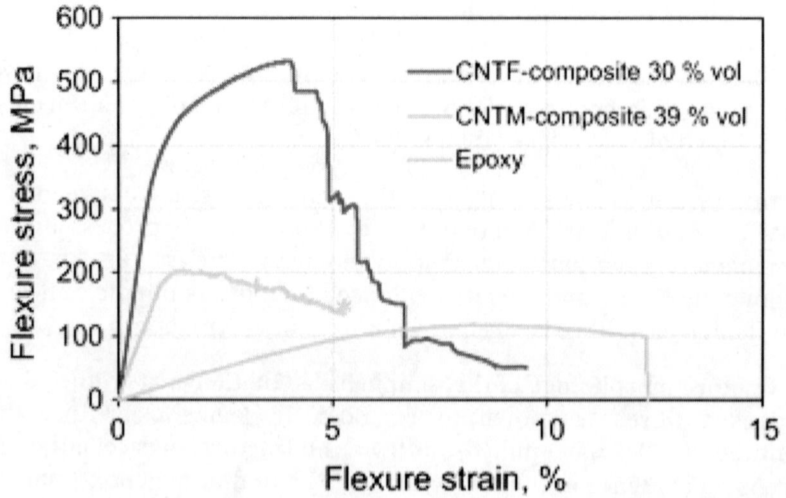

Figure 9. Representative bending stress–strain curves for pure epoxy matrix and nanocomposites. Note that the CNTM-composites test stopped due to reaching maximum rig deflection for the thinner sample used

Table 2. Flexural properties of the pure epoxy and CNT composites

Sample	Volume fraction (%)	Flexural strength		Flexural modulus		Maximum surface strain (%)
		MPa	GPa/SG (N/tex)	GPa	GPa/SG (N/tex)	
Pure epoxy	0	124.5 ± 4.5	0.116 ± 0.004	2.1 ± 0.1	2.0 ± 0.1	12.7 ± 1.3
CNTF-composite	30	514 ± 30	0.396 ± 0.023	47.3 ± 2.8	36.5 ± 2.2	4.5 ± 0.5
CNTM-composite[a]	39	185 ± 22	0.172 ± 0.021	17.0 ± 0.8	15.8 ± 0.7	>5

[a] Note that the CNTM-composites test stopped due to reaching maximum rig deflection for the thinner sample used

The CNT fibre composites with a 30 % volumetric loading showed increases in specific strength and stiffness by factors of ~3.4 and ~18, respectively. However, as with the tensile samples, the uncondensed CNT mat composites, even for a higher volumetric loading of 39 %, showed only minor enhancement in strength and stiffness with improvement factors of ~1.5 and ~8.

While in the tensile tests, the stress is transferred by shear from the gripped surfaces into the sample, in bending tests, stress transfer by shear provides the axial loading of the top (compression in this geometry) and bottom (tensile) surfaces, these stresses increasing from the mid plane towards each of the two surfaces.

Figure 10a, b is side views of the CNTF-composite which corresponds to the stress/strain curve of Fig. 9.

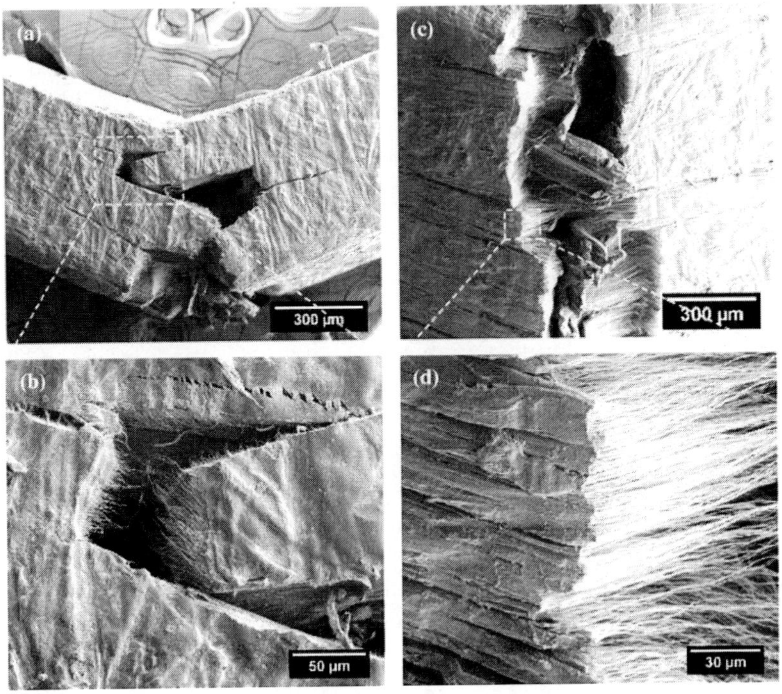

Figure 10. Bending-failure analysis of CNTF-composites: **a, b** side views showing delamination, **c** fracture at the tensile surface of the sample with an enlarged region **d** displaying pulled-out nanotube bundles bridging the crack which has opened to ~0.5 mm

Tensile failure initiated on the face of maximum tensile stress and progressed towards the compressive face. There is also evidence of a degree of delamination parallel to the fibre axis, although whether this path is between or within the original fibres is not clear. Examination of the compressive face gives no indication of local buckling. Figure 10c is the view normal to the tension surface, showing the crack which initiated failure. At higher magnification, Fig. 10d, pulled-out nanotube bundles bridging the crack are clearly evident.

Composites based on nanotube mats (CNTM) showed a very different failure mode in bending. Microscopic examination of the compression face (top of side view in Fig. 11a) showed buckling by the surface layers 'arching' away from the original surface, with interlamellar shear and localized buckling, even more obvious in higher magnification (Fig. 11b).

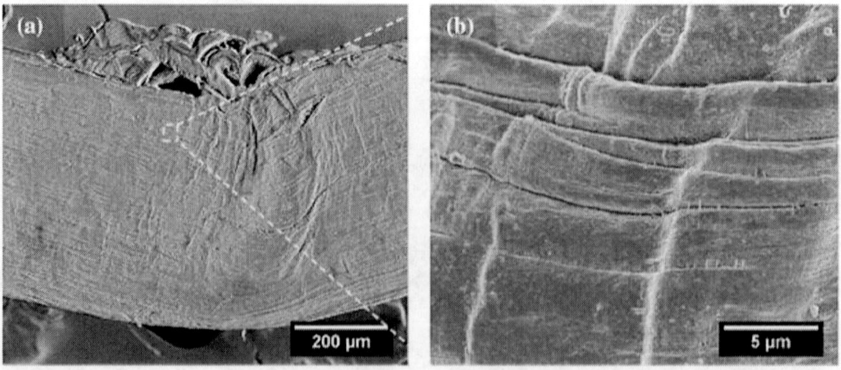

Figure 11. Bending-failure analysis of the CNTM-composite: **a** side view with distinct buckling where several layers were 'arching' away from the original surface under the loading, **b** enlarged image giving evidence for shear failures towards the compression side

The most striking aspect is that the CNTM-composites bear no evidence of tensile failure at all. In the bending rig with its maximum permissible strain of 5 %, the peak load was reached at around 1 % strain. However, the relatively thin samples (<0.5 mm) never fractured. The deformation mode is, thus, a combination of compression and shear, with no tensile drawing or fracture apparent. The modest improvement in strength compared to the pure epoxy simply represents the onset of plastic failure due to buckling at the compressive face at ~1 % strain. There is, however, a considerable improvement in stiffness of the material for 39 % volume fraction mats of a factor of 8. The inclusion of the CNTs in the epoxy appears to have greatly enhanced the plasticity of the material in bending as well as its stiffness.

The buckling observed at the compressive face of the bending samples is considerable, and much of the yield flow seen in force deflection curves is associated with this form of compressive failure. Basically, both fibre and mats have essentially zero compressive strength until incorporated in the resin. Here, the restraining influence of the matrix delays the onset of Euler buckling. However, the difference between the fibre and mat composite samples, we could not find any evidence of buckling in the fibre composite, demonstrates that the additional mechanical coherence between the bundles, as achieved during the condensation stage of the fibre production, contributes to their considerably greater resistance to compressive buckling. In addition, it is possible that the improved quality of bundle alignment achieved in the fibres in it adds to resistance to buckling.

Ethical Statement

This work was carried out with financial support of a Marie Curie Intra-European Human Mobility Fellowship Programme: "Project No. 251477: CNTF-composites" which is gratefully acknowledged. However, the contents reflect only the authors' views and not the views of the University or the European Commission. Alan Windle is a director of Q-Flo Ltd., in which he has a shareholding, and consults in the general field of Materials Science.

REFERENCES

1. Behabtu N, Young CC, Tsentalovich DE, Kleinerman O, Wang X, Ma AWK, Bengio EA, Waarbeek RF, Jong JJ, Hoogerwerf RE, Fairchild SB, Ferguson JB, Maruyama B, Kono J, Talmon Y, Cohen Y, Otto MJ, Pasquali M (2013) Strong, light, multifunctional fibers of carbon nanotubes with ultrahigh conductivity. Science 339(6116):182–186.

2. Jiang C, Saha A, Young CC, Hashim DP, Ramirez CE, Ajayan PM, Pasquali M, Marti AA (2014) Macroscopic nanotube fibers spun from single-walled carbon nanotube polyelectrolytes. ACS Nano 8(9):9107–9112.

3. Zhang M, Atkinson KR, Baughman RH (2004) Multifunctional carbon nanotube yarns by downsizing an ancient technology. Science 306(5700):1358–1361.

4. Vilatela JJ, Windle AH (2010) Yarn-like carbon nanotube fibers. Adv Mater 22(44):4959–4963.

5. Aleman B, Reguero V, Mas B, Vilatela JJ (2015) Strong carbon nanotube fibers by drawing inspiration from polymer fiber spinning. ACS Nano 9(7):7392–7398.

6. Xu F, Sadrzadeh A, Xu Z, Yakobson BI (2013) Can carbon nanotube fibers achieve the ultimate conductivity?—coupled-mode analysis for electron transport through the carbon nanotube contact. J Appl Phys 114:063714.

7. Lekawa-Raus A, Patmore J, Kurzepa L, Bulmer J, Koziol K (2014) Electrical properties of carbon nanotube based fibers and their future use in electrical wiring. Adv Funct Mater 24:3661–3682.

8. Koziol K, Vilatela J, Moisala A, Motta M, Cunniff P, Sennett M, Windle A (2007) High-performance carbon nanotube fiber. Science 318:1892–1895.

9. Windle A (2014) Carbon nanotube fibres: science and technology transfer. Nanoscience and technology for mankind. Humana Press, Allahabad

10. Bradford PD, Wang X, Zhao HB, Maria JP, Jia QX, Zhu YT (2010) A novel approach to fabricate high volume fraction nanocomposites with long aligned carbon nanotubes. Compos Sci Technol 70(13):1980–1985.

11. Mecklenburg M, Mizushima D, Ohtake N, Bauhofer W, Fiedler B, Schulte K (2015) On the manufacturing and electrical and mechanical properties of ultra-high wt% fraction aligned MWCNT and randomly oriented CNT epoxy composites. Carbon 91:275–290.

12. Mora RJ, Vilatela JJ, Windle AH (2009) Properties of composites of carbon nanotube fibres. Compos Sci Technol 69(10):1558–1563.

13. Vilatela JJ, Khare R, Windle AH (2012) The hierarchical structure and properties of multifunctional carbon nanotube fibre composites. Carbon 50(3):1227–1234.

14. Qiu J, Terrones J, Vilatela JJ, Vickers ME, Elliott JA, Windle AH (2013) Liquid infiltration into carbon nanotube fibers: effect on structure and electrical properties. ACS Nano 7(10):8412–8422.

15. Gspann TS, Montinaro N, Pantano A, Elliott JA, Windle AH (2015) Mechanical properties of carbon nanotube fibres: St

Venant's principle at the limit and the role of imperfections. Carbon 93:1021–1033.

16. Gspann TS, Small FR, Windle AH (2014) Spinning of carbon nanotube fibres using the floating catalyst high temperature route: purity issues and the critical role of sulphur. Faraday Discuss 173:47–65.

17. Ayres CE, Jha BS, Meredith H, Bowman JR, Bowlin GL, Henderson SC et al (2008) Measuring fiber alignment in electrospun scaffolds: a user's guide to the 2D fast Fourier transform approach. J Biomater Sci Polym Ed 19(5):603–621.

18. Stein IY, Wardle BL (2013) Coordination number model to quantify packing morphology of aligned nanowire arrays. Phys Chem Chem Phys 15:4033–4040.

19. Duesberg GS, Loa I, Burghard M, Syassen K, Roth S (2000) Polarized Raman spectroscopy on isolated single-wall carbon nanotubes. Phys Rev Lett 85(25):5436–5439.

20. Liu L, Zhang BM, Wang DF, Wu ZJ (2006) Effects of cure cycles on void content and mechanical properties of composite laminates. Compos Struct 73(3):303–309.

21. Grunenfelder LK, Centea T, Hubert P, Nutt SR (2013) Effect of room-temperature out-time on tow impregnation in an out-of-autoclave prepreg. Compos Part A Appl Sci 45:119–126.

22. Tavares SS, Michaud V, Manson JAE (2009) Through thickness air permeability of prepregs during cure. Compos Part A Appl Sci 40(10):1587–1596.

23. Huang H, Talreja R (2005) Effects of void geometry on elastic properties of unidirectional fiber reinforced composites. Compos Sci Technol 65(13):1964–1981.

24. Wardle BL, Saito DS, Garcia EJ, Hart AJ, de Villoria RG, Verploegen EA (2008) Fabrication and characterization of ultralight-volume-fraction aligned carbon nanotube-polymer composites. Adv Mater 20(14):2707–2714.

25. Centea T, Hubert P (2011) Measuring the impregnation of an out-of-autoclave prepreg by micro-CT. Compos Sci Technol 71(5):593–599.

26. Zhu HY, Li DH, Zhang DX, Wu BC, Chen YY (2009) Influence of voids on interlaminar shear strength of carbon/epoxy fabric laminates. Trans Nonferr Metal Soc 19:S470–S475

27. Ginga NJ, Chen W, Sitaraman SK (2014) Waviness reduces effective modulus of carbon nanotube forests by several orders of magnitude. Carbon 66:57–66.

28. Stein IY, Lewis DJ, Wardle BL (2015) Aligned carbon nanotube array stiffness from stochastic three-dimensional morphology. Nanoscale 7:19426–19431.

29. Fisher FT, Bradshaw RD, Brinson LC (2002) Effects of nanotube waviness on the modulus of nanotube-reinforced polymers. Appl Phys Lett 80(24):4647–4649.

30. Fisher FT, Bradshaw RD, Brinson LC (2003) Fiber waviness in nanotube-reinforced polymer composites—I: modulus predictions using effective nanotube propeties. Comp Sci Technol 63:1689–1703.

31. Rafiee R (2013) Influence of carbon waviness on the stiffness reduction of CNT/polymer composites. Comp Struct 97:304–309.

32. Varischetti J, Jang JS, Gibson RF, Suhr J (2013) Effect of filler waviness and orientation on the damping behavior of CNF-reinforced epoxy composites. J Mat Sci 48:832–840.

33. Cebeci H, de Villoria RG, Hart AJ, Wardle BL (2009) Multifunctional properties of high volume fraction aligned carbon nanotube polymer composites with controlled morphology. Compos Sci Technol 69(15–16):2649–2656.

34. Nam TH, Goto K, Nakayama H, Oshima K, Premalal V, Shimamura Y, Inoue Y, Naito K, Kobayashi S (2014) Effects of stretching on mechanical properties of aligned multi-walled carbon nanotube/epoxy composites. Compos Part A Appl Sci 64:194–202.

35. Lopes PE, van Hattum F, Pereira CMC, Novoa PJRO, Forero S, Hepp F, Pambaguian L (2010) High CNT content composites with CNT Buckypaper and epoxy resin matrix: impregnation behaviour composite production and characterization. Compos Struct 92(6):1291–1298.

36. Spitalsky Z, Tsoukleri G, Tasis D, Krontiras C, Georga SN, Galiotis C (2009) High volume fraction carbon nanotube–epoxy composites. Nanotechnology 20(40):405702.

37. Li M, Wang Z, Liu Q, Wang S, Gu Y, Li Y, Zhang Z (2015) Carbon nanotube film/epoxy composites with high strength and toughness. Polym Compos.

38. Cheng QF, Wang JP, Wen JJ, Liu CH, Jiang KL, Li QQ, Fan SS (2010) Carbon nanotube/epoxy composites fabricated by resin transfer molding. Carbon 48(1):260–266.

39. Boncel S, Sundaram RM, Windle AH, Koziol KKK (2011) Enhancement of the mechanical properties of directly spun CNT fibers by chemical treatment. ACS Nano 5(12):9339–9344.

40. Miller SG, Williams TS, Baker JS, Sola F, Lebion-Colon M, McCorkle LS, Wilmoth NG, Gaier J, Chen M, Meador MA (2014) Increased tensile strength of carbon nanotube yarns and sheets through chemical modification and electron beam irradiation. ACS Appl Mater Interfaces 6(9):6120–6126.

41. Filleter T, Espinosa HD (2013) Multi-scale mechanical improvement produced in carbon nanotube fibers by irradiation cross-linking. Carbon 56:1–11.

42. Naraghi M, Filleter T, Moravsky A, Locascio M, Loutfy RO, Espinosa HD (2010) A multiscale study of high performance double-walled nanotube-polymer fibers. ACS Nano 4(11):6463–6476.

43. Ma W, Liu L, Zhang Z, Yang R, Liu G, Zhang T, An X, Yi X, Ren Y, Niu Z, Li J, Dong H, Zhou W, Ajayan PM, Xie S (2009) High-strength composite fibers: realizing true potential of carbon nanotubes in polymer matrix through continuous reticulate architecture and molecular level couplings. Nano Lett 9(8):2855–2861.

44. Shimamura Y, Oshima K, Tohgo K, Fujii T, Shirasu K, Yamamoto G, Hashida T, Goto K, Ogasawara T, Naito K, Nakano T, Inoue Y (2014) Tensile mechanical properties of carbon nanotube/epoxy composite fabricated by pultrusion of carbon nanotube spun yarn preform. Compos A Appl Sci Manuf 62:32–38.

45. Lachman N, Wagner HD (2010) Correlation between interfacial molecular structure and mechanics in CNT/epoxy nanocomposites. Compos Part A Appl Sci 41(9):1093–1098.

46. Bucknall CB (1977) Toughened plastics. Applied Science Publishers Ltd, London, UK

3

STUDY OF THE WEAR AND CORROSION PERFORMANCE OF HARD COATINGS APPLIED BY DIFFERENT PROCESSES ON LOW CARBON STEEL

Carlos Roberto Camello Lima[1*], Rodolfo Libardi[1], Flávio Camargo[2], Miguel Angel Reyes Mojena[3], Hipólito Domingo Carvajal Fals[3], Valtair Antonio Ferraresi[4]

[1]Methodist University of Piracicaba, Santa Bárbara d'Oeste, São Paulo, Brazil

[2]Ogramac Surface Engineering, Santo Antônio de Posse, São Paulo, Brazil

[3]Oriente University, Santiago de Cuba, Cuba

[4]Federal University of Uberlândia, Uberlândia, Minas Gerais, Brazil

ABSTRACT

Most of materials used in applications that require high wear resistance also undergo corrosive action of the service environment. The damage caused by the combination of both processes on the materials leads to huge economic losses in industry. Hard coatings are widely used as engineering solution to increase the lifetime of components operating in conditions of wear combined with corrosion. Nowadays, the most versatile techniques for applying such coatings are welding and thermal spraying. This work evaluates and compares the properties of coatings obtained by thermal spraying and welding using selected feeding materials. The performance of the coatings in corrosion and erosive wear tests is discussed. From the erosion tests performed for impact angle of 90°, a similar performance for the thermally sprayed tungsten carbide coatings and the welded stainless steel coatings with higher input energy was observed. Welded samples presented the best performance in corrosion test.

Keywords: Hard Coatings, HVOF, Welding, Erosive Wear, Corrosion

1. INTRODUCTION

One of the leading causes for degradation and failure of equipment parts is wear in its diverse forms as abrasion or erosion [1] . Frequently, in several industries, the erosive environment presents corrosive agents making the search for failure prevention still more challenging. Tribo-corrosion is encountered in several technological areas causing damages to equipment parts and compromising safe operation [2] - [4] . As a less costly alternative to parts fully made of wear and corrosion resistant materials, coatings have become a very important element for equipment manufacture and repair. Equipment can achieve improved performance if properly coated. The recovery of pumps, valves and other parts of industrial machinery using engineered coatings constitutes today one of the most important engineering alternatives [5] [6] . Coatings are thick films and can be applied by several techniques like welding, physical or chemical vapor deposition, electroplating and thermal spraying [7] [8] . There are a number of different welding processes requiring the selection of the process (or processes) suitable for a given application [9] [10] . Normally elected for the application of coating overlays, Gas Metal Arc Welding (GMAW) uses heat from an electric arc between a naked electrode, fed continuously, and the base metal [11] . Problems like discontinuities, lack of fusion, lack of penetration, slag inclusions, cracks and pores can occur [12] [13] .

Thermal spray processes are very suitable to obtain coatings of high performance for protection or repair of components [14] [15] . All thermal spray processes use a gas generator as kinetic component. The balance between heat energy and kinetic energy can vary in large proportions, which results in coatings of different properties depending on the technique used for thermal spraying [16] . The combination of a structurally adequate substrate material and a proper coating has proven to be a highly flexible and economical manufacture strategy, mainly for the use in highly aggressive environments [7] [14] . Coatings using WCCo alloys, deposited by the thermal spray process, have been widely employed in applications against wear and corrosion by combining several advantages such as high hardness, resistance to abrasive and erosive wear, and corrosion resistance in various media. Cr_3C_2-NiCr coatings offer superior corrosion and oxidation resistance and have a high melting point. In operating conditions, they also retain high hardness, strength and wear resistance up to temperatures around 900°C [17] . There is technical feasibility of using coatings of WCCo composites deposited by HVOF (High Velocity Oxygen Fuel) due to the excellent mechanical properties, wear and corrosion resistance, high deposition rate and superior surface finish compared to other processes deposition. By this process, the coatings have extremely low porosity (typically < 1%) and high adhesion strength compared to conventional spraying processes. The coatings deposited by HVOF thermal spraying confer an excellent ability to protect the substrate [18] . The aim of this work is to evaluate and compare the wear and corrosion properties of coatings obtained by thermal spraying and welding using selected feeding materials.

2. MATERIALS AND METHODS

The substrate material used in this work was an AISI 1020 low carbon steel. The materials for application in the coatings were chosen looking for a combined resistance to erosive wear as well as corrosive environments. Commercial powders of WC17Co (Praxair, Concord, NH, USA), −45 μm ± 15 μm and Cr_3C_2-25NiCr (Praxair, Concord, NH, USA), −45 μm ± 15 μm were applied by HVOF. A stainless steel wire AWS 308LSi (BELGO, São Paulo, BR), 1.6 mm diameter, was used for GMAW coating deposition. The chemical composition of the applied substrate and feedstock materials are presented in Table 1. GMAW process with inert gas (MIG) was used for welding overlay. The power source used was an adapted MTE DIGITEC 600 with double wire (IMC/ LABSOLDA, Uberlândia, MG, BR). The three distinctive sets of process parameters used for welded coating deposition are presented in Table 2. For HVOF spraying, the equipment used was a TAFA-JP 5000 (Praxair, Concord, NH, USA) with the parameters listed in Table 3.

Microhardness measurements were performed on Shimadzu computerized equipment with a capacity of 0.01 kg to 2 kg in Vickers scale. All the welded samples were obtained from deposits and zones of interest in the form of prismatic bodies of 10 × 8 × 4 mm. Because of the high roughness of the as welded coatings, all the welded samples were grinded and polished with sand paper to 1200 mesh while the sprayed coatings were kept in the as sprayed condition in order to evaluate the samples without additional grinding and polishing work. Surface roughness measurements were carried out through a Mitutoyo digital Roughness TIME TR-200, with precision of 0.001 microns and processed through the interface for TIMESURF TR 200 V1.4. The results are presented as a mean of ten measurements. Surface roughness of the obtained coatings measured in the as sprayed condition for thermal sprayed samples and after polishing for welded samples. For image analysis and microstructure evaluation Scanning Electron Microscopy was applied (JEOL-JXA-840 and HITACHI TM 3000).

Table 1. Chemical composition of used materials (wt.%).

Material	C	Cr	Mn	Si	Co	Fe	Ni	W	Other
AISI 1020	0.25	0.0	0.6	0.35	0.0	Bal.	0.0	0.0	0.1
308LSi	0.03	19.4	0.9	0.8	0.0	68.7	9.9	0.0	0.002
WC17Co	5.3	0.0	0.0	0.0	17.0	0.0	0.0	Bal.	0.0
$Cr_3C_2$25NiCr	11.0	Bal.	0.0	0.0	0.0	0.0	19.0	0.0	0.002

Table 2. Welding parameters for GMAW coating deposition.

Sample	V weld (cm/min)	V feed (m/min)	Voltage (V)	Energy (Kj/cm)	Feeding material	Dilution (%)
1	80.0	14.0	30.0	0.212	ER308LSi	29.06
2	60.0	12.0	30.0	0.255	ER308LSi	35.02
3	80.0	12.0	30.0	0.192	ER308LSi	30.00

Table 3. HP-HVOF deposition parameters.

Feedstock material	Oxygen		Kerosene		Air		Feed rate (g/min)	Stand off distance (mm)
	Pressure (bar)	Flow rate (SLPM)	Pressure (bar)	Flow rate (SLPM)	Pressure (bar)	Flow rate (SLPM))		
WC17Co	10.0	944	8.6	0.378	3.4	12	76	350
$Cr_3C_2$25NiCr	10.3	873	6.9	0.378	3.4	12	38	250

Erosion tests were accomplished in an adapted mass testing machine designed by Erosive Welding Lab, Federal University of Uberlândia, BR. The testing machine was designed based on similar equipment developed by Desale et al. [19] and incorporating some details and procedures of the ASTM G73-10 [20] , but did not specifically following that standard. The scheme of the erosion testing machine is presented in Figure 1. The samples for testing were obtained from deposits and zones of interest in the form of prismatic bodies of 10 × 8 × 4 mm. All the welded samples were polished with sand paper to 1200 mesh in order to reduce the characteristic surface irregularities. For pure erosion testing the abrasive was a mixture of 4 liters of water and abrasive particle concentration of 400 g/liter.

The used abrasive were particles of ferric oxide, HRC = 40 - 50, 600 um diameter. The design of the specimen allows evaluating the impact angle of abrasive particles at 90° and 30°. In this work, only 90° impact angle was tested. The impact velocity of the abrasive particles was calculated as v = 31 m/s. As mentioned above trials were performed for each condition at a time of 2 hours, at intervals of 20 minutes, during which wear rates were determined. The wear was determined by difference in weight before and after test employing a precision analytical balance (0.1 mg). Each test was repeated under the same conditions 3 times getting the average value results for each sample.

The polarization curves for the different materials tested were obtained in a potentiostat AUTOLAB PGSTAT 302N (Metrohm Autolab, Utrecht, NL) connected to an interface and GPES professional package, which allows the choice of the electrochemical technique to use, graphical options and determination of the main parameters test (corrosion potential sting passivation current, corrosion rate, etc.). As corrosive medium NaOH was used in concentration of 1 mol (PH = 14). For measurements, a platinum counter electrode and a reference electrode of Ag/AgCl were employed to a scanning speed of 1.66 mV/s from –800 to 900 mV with respect to corrosion potential.

3. RESULTS AND DISCUSSION

3.1. Mechanical Properties

Table 4 shows the average surface roughness and microhardness results for the obtained coatings. From the analysis of Table 4, it can be observed that a coating with high hardness was achieved by using the HVOF spraying in the case of WC17Co, 1042 HV, about 62% higher than the other HVOF coating of $Cr_3C_2$25NiCr.

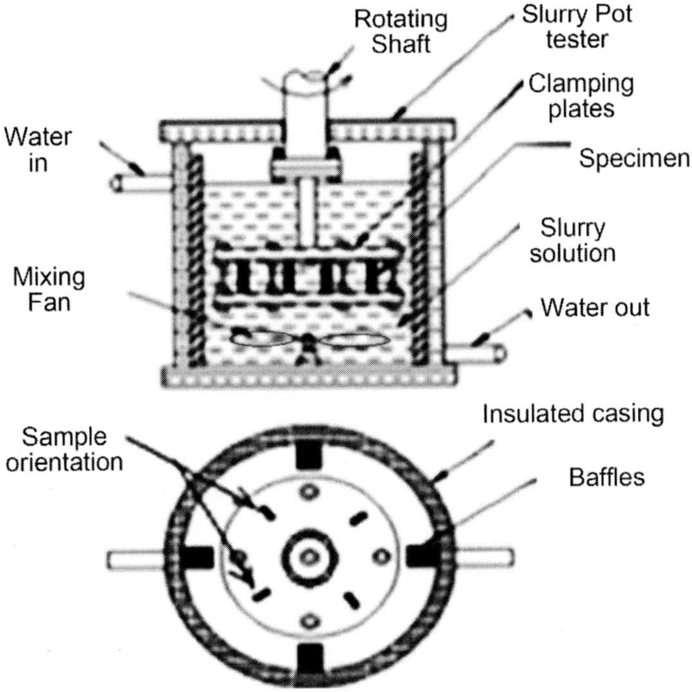

Figure 1. Schematic representation of the erosive wear test.

Table 4. Surface roughness and microhardness results for thermal sprayed and welded coatings.

Sample	Roughness-Ra (μm)	Microhardness $HV_{0.5}$
ER308LSi 1	0.86	386
ER308LSi 2	0.96	404
ER308LSi 3	0.85	342
WC-17Co	3.12	1042
Cr_3C_2-25NiCr	5.44	692

Note: Results are presented as the mean value of 10 measurements with a maximum standard deviation of 10%.

The microhardness of the three ER308LSi stainless steel welding coatings was around 36% of the maximum value for WC17Co alloy. The welded material is basically constituted of iron with almost 20% of chromium in its composition, which contributed to a high hardness. $Cr_3C_2$25NiCr that has higher Cr content (almost 70%) has presented the second higher hardness. WC17Co has essentially tungsten in its composition with a high content of WC, which also contributes to that high hardness. It should be observed that the roughness of HVOF coatings was measured in the as sprayed condition while GMAW welded coating was machined and abraded before measurement. $Cr_3C_2$25NiCr coatings have presented the highest surface roughness in comparison with the other coatings. The typical microstructure of the obtained coatings can be seen in Figure 2 for welded and HVOF sprayed samples. Figure 3 shows the cross sectional images of the welded (typical) and thermally sprayed coatings.

It can be observed that, as expected, a typical casting structure results for welded coatings. Since the tested samples were obtained from the center of the welded beads, that includes the fusion zone and the heat affected zone, as showed in Figure 3(a) and Figure 3(d), a very homogeneous microstructure is exposed to the erosion and corrosion tests. From the coatings applied by HVOF spraying, the lamellar structure is dominant and presents certain level of pores and oxides, as presented in the enlarged detail of Figure 3(b) and Figure 3(c) for $Cr_3C_2$25NiCr and WC17Co, respectively. It can be also observed differences in carbide size and distribution for $Cr_3C_2$25NiCr and WC17Co. Tungsten carbide coatings (Figure 3(e)) present a smaller carbide size than Chromium carbide (Figure 3(f)) and then a higher amount of carbides in the microstructure.

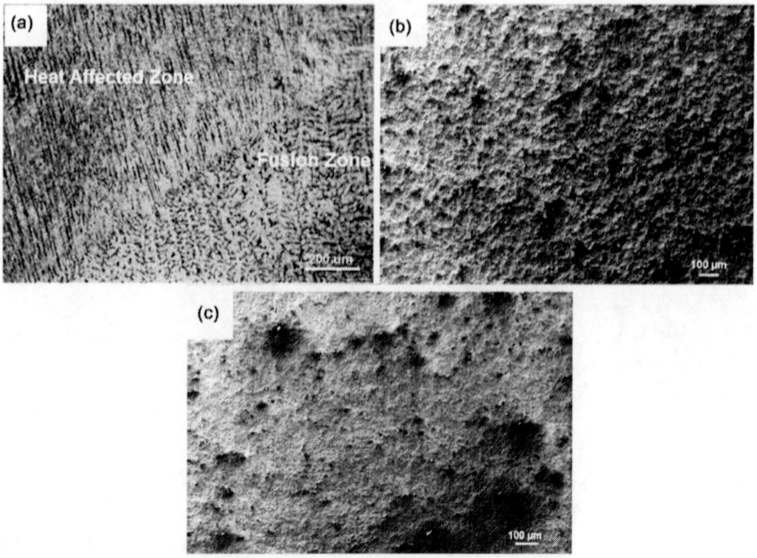

Figure 2. SEM surface images of the coatings of ER308LSi obtained by welding showing (a) the typical welded microstructure of samples, and the HVOF sprayed coatings of (b) WC17Co and (c) $Cr_3C_2$25NiCr.

3.2. Corrosion Results

The results of the polarization tests can be seen in Figure 4. The result for the substrate alone is also presented for comparison. The Tafel slope analysis of each one of the coatings is detailed inTable 5.

In a general way, all the coatings have presented a very low oxidation rate that is probably related to the characteristic impedance of the coatings, as defined by Toma et al. [21] . From the analysis of the results showed in Figure 4 and Table 5, it can be observed that the best performance was that of the ER308LSi samples, followed by WC17Co. Comparing the corrosion results for the three welded samples, it is possible to observe a variation of two orders of magnitude among the samples that could be credited to the differences in dilution and formation of carbides.

The worst result was for the Cr_3C_2-25NiCr sample, with a corrosion rate much higher than the welded samples and about twice the WCCo sample. It has to be stressed that the corrosion results of the thermally sprayed samples are even worse than the results for the carbon steel substrate. These results can be justified by the corrosion resistance and passivity of the respective binder in the carbide sprayed coatings. As stated by Souza and Neville [22] , corrosion proceeds primarily by dissolution of the Co or NiCr binder phase.

It could be expected a better corrosion performance for the Cr_3C_2-25NiCr that has about 20% of Cr when compared to carbon steel. Then, an additional explanation for such results can be the differences in the surface roughness of these coatings, exposing a much larger surface area for the corrosion agent, as discussed by Çelik et al. [23] . It has to be stressed that the original roughness of the Cr_3C_2-25NiCr coating was kept in the as sprayed condition exactly to allow this comparison.

3.3. Erosive Wear Results

The erosive wear results are presented in Figure 5 for an impact angle of 90°. Thermally sprayed samples were tested with initial surface roughness values corresponding to that of the as deposited coating, i.e., 3.12 μm and 5.44 μm for WC17Co and $Cr_3C_2$25NiCr, respectively. Because of the extremely high roughness characteristic of the welded

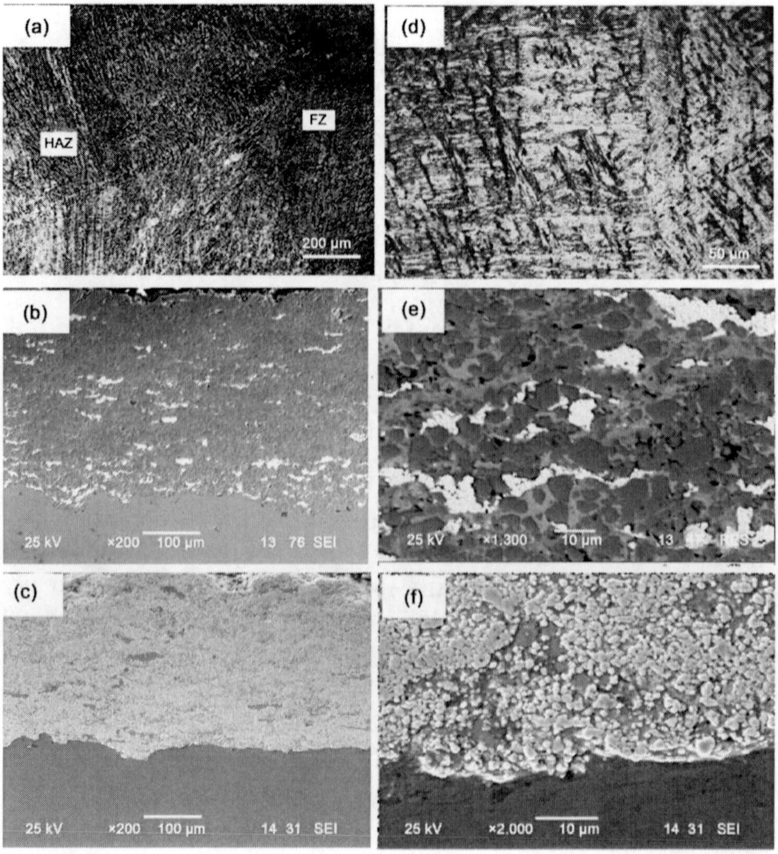

Figure 3. SEM cross sectional typical images of the (a) welded ER308LSi, (b) $Cr_3C_2$25NiCr and (c) WC17Co thermally sprayed coatings showing the typical casting and lamellar structures. Detailed microstructures are observed in images (d), (e) and (f) for the coatings on the left, respectively. Larger carbide size can be observed for $Cr_3C_2$25NiCr (e) in comparison with WC17Co (f).

deposits, the ER308LSi tested samples were grinded and polished to 1200 mesh, reaching a much lower roughness around 1 µm. Observing the microstructure of the coatings (Figure 3), it is possible to infer an analogy between the structure and behavior in erosive wear.

A regular homogeneity in the distribution of the carbides in the matrix of cobalt can be seen in the WC17Co (Figure 3(f)) as well as higher hardness may explain its high wear resistance. Differences in carbide sizes could also explain the higher resistance, as evidenced by Yang et al.

[24] who show an increased rate of wear with increasing grain size of carbides. In the case of the $Cr_3C_2$25NiCr coating, a structure with relatively high porosity and low carbide homogeneity distribution (Figure 3(e)) associated with large carbide size can justify its performance in wear tests.

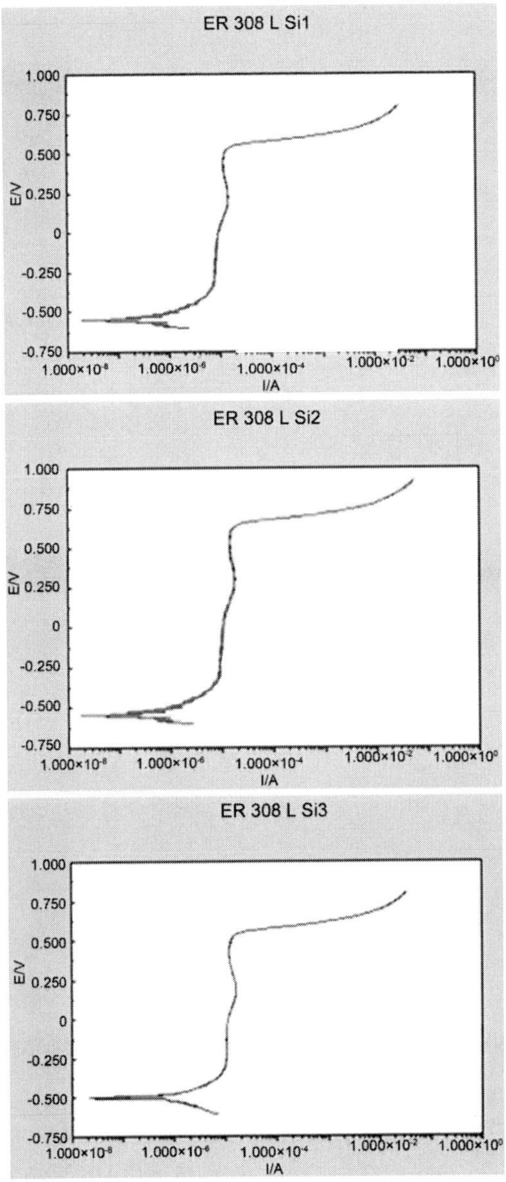

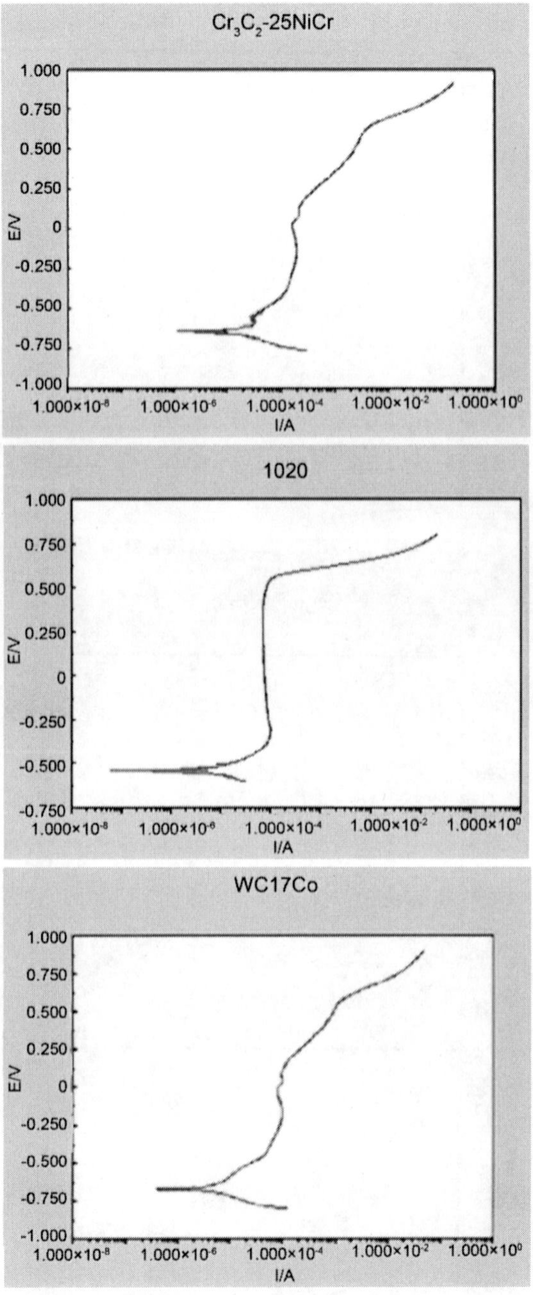

Figure 4. Corrosion results from polarization tests for the tree welded ER308LSi samples, the AISI 1020 substrate and the two HVOF sprayed samples.

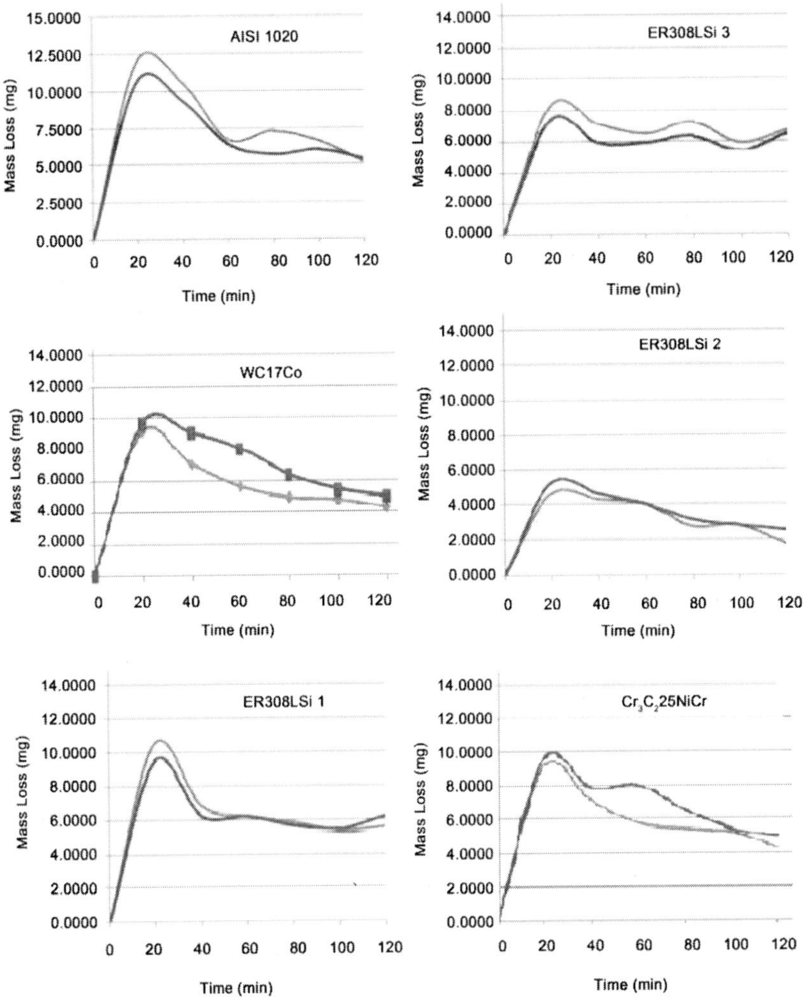

Figure 5. Mass loss for the AISI 1020 substrate, ER308LSi, WC17Co and $Cr_3C_225NiCr$ coated samples in the erosive wear tests.

The analysis of the worn surface (Figure 6) showed a high degree of detachment of particles with formation of craters and subsequent destruction of these craters and borders (Figure 6(d)), then justifying the low wear resistance. In all cases the stabilizing period is characterized by an intense wear of the coating, and this has been reflected by the literature [25] [26] . From the analysis of the eroded surfaces, it could be observed that the mechanism dominating the erosion process is brittle fracture type whose value is maximum for normal impact angle.

Table 5. Corrosion test results of polarization curves from Tafel slope analysis.

Parameters	Substrate	Thermally sprayed coatings		Welded coatings		
	AISI 1020	WC-17Co	Cr3C2-25NiCr	ER308SLi 1	ER308SLi 2	ER308SLi 3
I_{corr} (A)	4.026E−7	4.737E−6	1.288E−5	2.588E−7	2.590E−7	1.198E−7
L_{corr} (A/cm^2)	5.033E−7	5.921E−6	1.61E−5	3.234E−7	3.866E−7	1.498E−7
R_p (Ohm)	1.352E+4	3.046E+3	3.863E+2	5.690E+3	1.043E+4	1.251E+2
$E_{corr,\ obs}$ (V)	−0.475	−0.677	−0.530	−0.512	−0.538	−0.491
$E_{corr,\ cal}$ (V)	−0.483	−0.683	−0.502	−0.509	−0.545	−0.497
Corr Rate (mm/year)	5.370E−4	2.800E−3	6.737E−3	3.451E−4	2.314E−4	1.598E−4

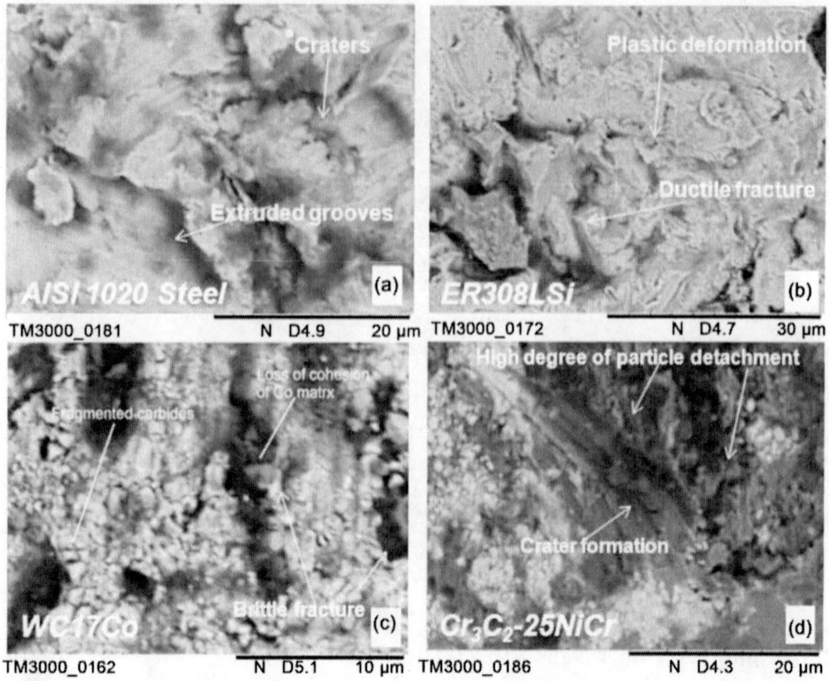

Figure 6. SEM images of the eroded surfaces of (a) AISI 1020 substrate, (b) ER308LSi, (c) WC17Co and (d) $Cr_3C_2 25NiCr$ coated samples.

The erosive wear of the welded coatings is preceded by fast surface hardening as evidenced by a previous comparison of systematic microhardness measurements in the eroded coatings throughout the coating thickness, from the substrate to the top surface, compared to the measurements of the as coated samples. The damage mechanism is ductile fracture characteristic of such materials where material detachment is preceded by plastic deformation, hardening and then loosening of wear particles as result of certain number of cycles, as stated previously by several authors [27] - [29] .

Figure 7 shows a comparative result for the volumetric wear of the tested samples in erosion. The calculation was performed by dividing the average of accumulated mass loss of the samples by the density of each material.

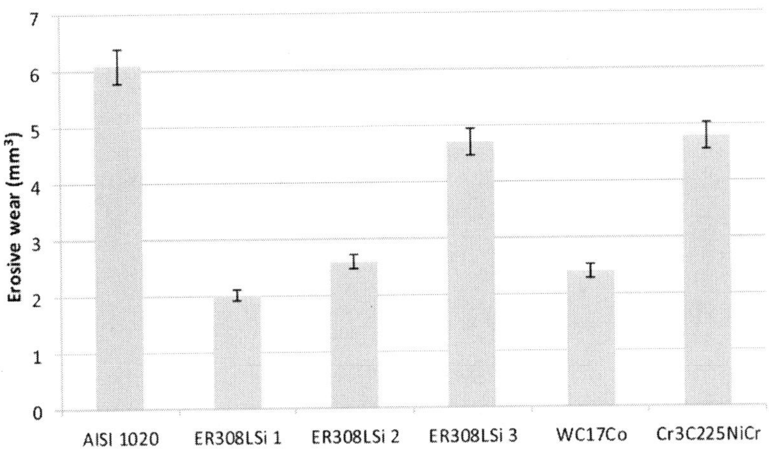

Figure 7. Volumetric erosive wear results of tested materials (t = 2 h, impact angle 90°).

The densities were obtained from the manufacturers of the materials: 7.85 g/cm^3, 7.9 g/cm^3, 4.6 g/cm^3 and 7.1 g/cm^3 for AISI 1020, ER308LSi, WC-17Co and Cr_3C_2-25NiCr respectively.

The best performance is that of the sample ER308LSi1. The behavior is very similar for the WC17Co applied coating that shows a slightly better result compared to the second stainless steel sample. $Cr_3C_2$25NiCr has presented the lowest erosive wear response for the coated samples, just compared to welded ER308LSi3. However, considering that the tests were performed only for impact angle of 90°, the results for 30° of impact angle have to be determined in order to fully evaluate the erosion performance of the applied materials.

4. CONCLUSIONS

Tungsten and chromium carbides alloys and stainless steel were deposited onto low carbon steel by thermal spraying or welding processes. The obtained coatings have shown a smooth microstructure without relevant level of defects. The best corrosion performance in polarization tests was for ER308LSi welded coatings, followed by WC17Co HVOF applied coatings.

The lowest corrosion resistance was obtained for Cr_3C_2-25NiCr sample that had performed even worse than the uncoated carbon steel. This not expected result was mainly credited to the huge difference in roughness of the thermal sprayed coatings compared to the polished welded coatings.

From the erosion tests performed for impact angle of 90°, it was found that the erosion resistance of the ER308LSi welded sample was the best one, followed quite closely by WC17Co HVOF sprayed sample, which had performed better than the other two welded samples.

ACKNOWLEDGEMENTS

This study was conducted with support from CAPES, the Brazilian government entity dedicated to training human resources, Project 131/11.

REFERENCES

1. Pamuk, U., Baydogan, M., Nilüfer, B. and Cimenoglu, H. (2000) An Energetic Approach to Abrasive Wear of a Martensitic Stainless Steel. Scripta Materialia, 42, 881-885.

2. Wood, R.J.K. (2007) Tribo-Corrosion of Coating: A Review. Journal of Physics D: Applied Physics, 40, 5502-5521.

3. Fedrizzi, L., Rossi, S., Bellei, F. and Deflorian, F. (2002) Wear-Corrosion Mechanism of Hard Chromium Coatings. Wear, 253, 1173-1181.

4. Gant, A.J., Gee, M.G. and May, A.T. (2004) The Evaluation of Tribo-Corrosion Synergy for WC-Co Hardmetals in Low Stress Abrasion. Wear, 256, 500-516.

5. Hattori, S., Taruya, K., Kikuta, K. and Tomaru, H. (2013) Cavitation Erosion of Silver Plated Coatings Considering Thermodynamic Effect. Wear, 300, 136-142.

6. Zhao, H.Y., Liu, S.G., Feng, Y.S., Yao, J.P. and Xu, Y.S. (2013) Failure Analysis of Desulfurization Slurry Circulating Pump and HVOF Coating Protection Technology Progress. Applied Mechanics and Materials, 320, 374-382.

7. Pawlowski, L. (1995) The Science and Technology of Thermal Spray Coatings. John Wiley & Sons, Chichester.

8. Davis, J.R., Ed. (2005) Handbook of Thermal Spray Technology. ASM International, Metals Park.

9. Davis, J.R. (1993) Hardfacing, Weld Cladding, and Dissimilar Metal Joining: Welding, Brazing and Soldering, Metals Handbook, Volume 6. 10th Edition, ASM International, Metals Park, 789-829.

10. Jeffus, L. (2004) Welding: Principles and Applications. 5th Edition, Delmar Learning, Clifton Park.

11. Naffakh, H., Shamanian, M. and Ashrafizadeh, F. (2008) Influence of Artificial Aging on Microstructure and Mechanical Properties of Dissimilar Welds between 310 Stainless Steel and INCONEL 657. Metallurgical and Materials Transactions A, 39, 2403-2415.

12. Davi, S.A. and Debroy, T. (1992) Current Issues and Problems in Welding Science. Science, 257, 497-502.

13. Hatala, M., Orlovsky, I. and Radchenko, S. (2014) Influence of Welding Parameters to Quality of Welds from Structural Steel. Key Engineering Materials, 581, 287-291.

14. Ernest, P. and Barbezat, G. (2008) Thermal Spray Applications in Powertrain Contribute to Saving of Energy and Material Resources. Surface and Coatings Technology, 202, 4428-4431.

15. Zhao, L. and Lugscheider, E. (2003) Influence of the Spraying Processes on the Properties of 316L Stainless Steel Coatings. Surface and Coatings Technology, 162, 6-10.

16. Guilemany, J.M., Nutting, J., Dong, Z. and De Paco, J.M. (1995) The Influence of WC-Co HVOF Thermal Spraying on the Microstructure of an Al-4%Cu Alloy Substrate. Scripta Metallurgica et Materialia, 33, 1055-1062.

17. Blatchford, M.T. (2001) Improvements in HVOF Sprayed Cermet Coatings Produced from SHS Powders. In: Berndt, C.C., Khor, K.A. and Lugscheider, E.F., Eds., Thermal Spray 2001: New Surfaces for a New Millennium, ASM International, Materials Park, 221-230.

18. Sobolev, V., Guilemany, J.M. and Nutting, J. (2004) High Velocity Oxy-Fuel Spraying. Maney, London.

19. Desale, G.R., Gandhi, B.K. and Jain, S.C. (2005) Improvement in the Design of a Pot Tester to Simulate Erosion Wear Due to Solid-Liquid Mixture. Wear, 259, 196-202.

20. (2010) ASTM G73-10, Standard Test Method for Liquid Impingement Erosion Using Rotating Apparatus. In: Bailey, S.J., Baldini, N.C. and Emer, S., Eds., Annual Book of ASTM Standards, American Society for Testing and Materials, Philadelphia.

21. Toma, D., Brandl, W. and Marginean, G. (2001) Wear and Corrosion Behaviour of Thermally Sprayed Cermet Coatings. Surface and Coatings Technology, 138, 149-158.

22. Souza, V.A.D. and Neville, A. (2005) Corrosion and Synergy in a WC-Co-Cr HVOF Thermal Spray Coating—Understanding Their Role in Erosion-Corrosion Degradation. Wear, 259, 171-180.

23. Celik, E., Sengil, I.A. and Avci, E. (1997) Effects of Some Parameters on Corrosion Behaviour of Plasma-Sprayed Coatings. Surface and Coatings Technology, 97, 355-360.

24. Yang, Q., Senda, T. and Ohmori, A. (2003) Effect of Carbide Grain Size on Microstructure and Sliding Wear Behavior of HVOF-Sprayed WC-12% Co Coatings. Wear, 254, 23-34.

25. Murthy, J.K.N., Rao, D.S. and Venkataraman, B. (2001) Effect of Grinding on the Erosion Behaviour of a WC-Co-Cr Coating Deposited by HVOF and Detonation Gun Spray Processes. Wear, 249, 592-600.

26. Murthy, J.K.N. and Venkataraman, B. (2006) Abrasive Wear Behaviour of WC-CoCr and Cr3C2-20(NiCr) Deposited by HVOF and Detonation Spray Processes. Surface and Coatings Technology, 200, 2642-2652.

27. Madsen, B.W. (1988) Measurement of Erosion-Corrosion Synergism with Slurry Wear Test Apparatus. Wear, 123, 127-142.

28. Zheng, Y.G., Yao, Z.M., Wei, X.Y. and Ke, W. (1995) The Synergistic Effect between Erosion and Corrosion in Acidic Slurry Medium. Wear, 186-187, 555-561.

29. Yao, Z.M., Zheng, Y.G. and Ke, W. (1995) The Influence of Applied Potential on the Erosion-Corrosion Behavior of AISI321 Stainless Steel in Acidic Slurry Medium. Wear, 186-187, 568-572.

4

NATURAL RUBBER BASED COMPOSITES COMPRISING DIFFERENT TYPES OF CARBON-SILICA HYBRID FILLERS. COMPARATIVE STUDY ON THEIR ELECTRIC, DIELECTRIC AND MICROWAVE PROPERTIES, AND POSSIBLE APPLICATIONS

Ahmed A. Al-Ghamdi[1], Omar A. Al-Hartomy[1], Falleh R. Al-Solamy[2], Nikolay Dishovsky[3], Mihail Mihaylov[3], Petrunka Malinova[3], Nikolay Atanasov[4,5]

[1]Department of Physics, Faculty of Science, King Abdulaziz University, Jeddah, KSA

[2]Department of Mathematics, Faculty of Science, King Abdulaziz University, Jeddah, KSA

[3]Department of Polymer Engineering, University of Chemical Technology and Metallurgy, Sofia, Bulgaria

[4]Department of Telecommunications, University of Telecommunications and Post, Sofia, Bulgaria

[5]Department of Communication and Computer Engineering, South-West University "Neofit Rilski" of Blagoevgrad, Blagoevgrad, Bulgaria

ABSTRACT

The paper presents a comparative study on the electric, dielectric and microwave properties of natural rubber based composites comprising dual phase fillers prepared from furnace carbon black or conductive carbon black with a different amount of silica. It has been established that, the specifics of the carbon phase have a marked strong effect upon the properties mentioned above. The interpenetration of the two filler phases and the grade of isolation of the conductive carbon phase by the dielectric one depend on the ratio between them. On the other hand, that leads to a change in all properties of the studied composites, which allows tailoring those characteristics.

Keywords: Polymer Composites, Hybrid Fillers, Dielectric Properties, Microwave Properties

1. INTRODUCTION

In the recent decades the technological progress has been causing a new type of contamination, namely, electromagnetic interference (EMI). EMI is an electromagnetic signal emitted by an electronic source which disturbs the performance of other similar devices in its vicinity [1] - [4] . Therefore, microwave absorbing materials have been finding wider application in modern technologies such as electronics and telecommunications [5] . The usage of such absorbers aims at reducing the radiation reflection, and ensuring EMI shielding and protection [5] - [9] . Being elastic, easy to process, possessing good mechanical properties and highly resistant to various natural factors, elastomers have been finding wider application in the design of microwave absorbing materials. However, most of the neat elastomers are dielectrics and absolutely transparent to electromagnetic radiation. For that reason, the elastomer based microwave absorbing materials consist of a rubber matrix and electromagnetic wave absorbents. The necessary electromagnetic performances of a rubber absorbing material (RAM) are owing to the electromagnetic wave absorbents while the rubber is their carrier [10] [11] .

An essential requirement for a good microwave absorbing material is the so-called “impedance matching”, which can decrease the reflection of the incident wave on the interface between free spaces and absorbing materials and be in favour of the improvement of absorbing properties. Dielectric and magnetic losses are also important parameters determining the absorbing properties of microwave absorbents [12] . As known, the main principles for developing elastomer based microwave absorbers are: finding a suitable dielectric rubber matrix and filler or a system of fillers possessing high dielectric and/or magnetic loss values (such as conductive carbon black, γ-Fe_2O_3, Ni-Zn-Fe_2O_4, Fe_3O_4, $SrFe_{12}O_{19}$, etc.) [13] - [18] . The crucial factor to obtain good microwave absorbers is the ability of the dielectric matrix or of any other dielectric phase to isolate completely the particles of the conducting filler, i.e. its particles to get in fewer contacts with each other [19] .

On the other hand, the constant improvement in wireless technologies and communications in recent years has set the necessity of creating antennae of miniature size, simplified structure and wider broadband. The compounds being used currently are copper coated laminates with high dielectric constant, which in fact are ceramic-polymer composites [20] . The shortcomings of the high dielectric ceramics are their being of high density and friability as well as hard process. Those shortcomings are overcome using various polymers [21] [22] . In this connection, Jin and co-workers have studied the properties of PTFE based composites, comprising $SrTiO_3$ of different amounts. The authors have established that, composites comprising about 60 wt% of $SrTiO_3$, whose dielectric constant is 12 ($\varepsilon_r' = 12$), are suitable for microwave radio frequency antenna applications [22] . In wideband communication systems, the antenna is the sole interfacing device attached to both the transmitter and receiver. It has been found that, the impedance bandwidth is of crucial importance for a high performance communication system. The operative frequency of an antenna is affected by dielectric permittivity which is influenced by the types of materials that it is made of. Antennas require a dielectric constant that is high enough (reflection loss lower than −10 dB in wide frequency range) for sufficient coupling to the excitation source in order to operate efficiently [23] .

Currently dual-phase furnace carbon black/silica fillers find wider application in rubber industry [24] [25] . According to our working hypothesis, if the silica phase (dielectric phase) in the hybrid fillers is located over the surface or penetrates the inner domains of carbon black aggregates (conductive phase), it should isolate them and restrict their interaction. Thus, the electric (conductivity), dielectric (dielectric constant) and microwave (impedance matching, reflection, absorption, etc.) properties of the composites comprising hybrid fillers could be controlled by varying the ratio between their conductive and dielectric phases. We have chosen to use impregnation technology for the preparation of such fillers since it is an affordable method facilitating an easy control over the quantitative ratio between the dielectric and conductive phases in the hybrid fillers.

The paper aims at comparing the electric, dielectric and microwave properties of natural rubber based composites comprising dual phase fillers prepared from furnace carbon black and conductive carbon black with a different amount of silicon dioxide (silica) and at evaluating the effect that the specifics of carbon phase have upon the composites properties. Another aim is to assess the possibilities for their application as microwave absorbers, shielding materials as well as for microwave radio frequency antennae.

2. EXPERIMENTAL

2.1. Materials

Natural rubber SVR 10 supplied by Hong Thanh Rubber Pty. Ltd. was used as a polymer matrix. The other ingredients such as zinc oxide (ZnO), stearic acid, N-tert-butyl-2-benzothiazole sulfenamide (TBBS) and sulphur (S) were commercial grades and used without further purification.

2.2. Hybrid Fillers Preparation

The hybrid fillers studied were prepared by modification with silicasol (40% of silica, pH-9 and density of 1.3 g/cm^3) of either industrial furnace carbon black type N 330 with specific surface area (BET) 75 m^2/g or conductive Printex XE-2B carbon black with specific surface area (BET) 1000 m^2/g (both produced by Orion Engineered Carbons GmbH). A mixture of 100 g of N 330 or Printex XE-2B carbon black and the needed amount of silicasol, corresponding to 3% or 7% of silica, was placed into a ball mill and 1.6 L of distilled water was poured over it. The mixture was impregnated for 2 hours. The product was placed into a drying chamber at 150°C and stayed till its complete drying. After that it was ground again in a ball mill for 2 hours. Then it was thermally treated at 440°C under 10^{-2} mm Hg vacuum for 2 hours in a reactor designed especially for the purpose. Thus, the thermal activation could be run at higher temperature without a negative effect upon the carbon black. Being removed from the reactor the product was ground again in a ball mill and was ready for further investigations. Four hybrid fillers based on furnace carbon black N 330 and Printex XE-2B conductive black, and 3% or 7% of silica were prepared according to the above method. The fillers were denoted as N 330/3, N 330/7, Pr/Si 3 and Pr/Si 7, respectively.

2.3. Preparation of Rubber Composites

The rubber compounds studied were prepared on a two-roll laboratory mill (rolls length/diameter 320 × 160 mm). Table 1 presents the compositions of the rubber compounds studied. The vulcanization of the natural rubber based compounds was carried out on an electrically heated hydraulic press using a special homemade mold at 150°C and 10 MPa.

2.4. Measurements

2.4.1. Electrical Properties Measurement

1) Volume Resistivity

Volume resistivity (ρ_v, Ω·m) of the studied composites was measured using two electrodes (2-terminal method) and calculated by the Equation (1):

$$\rho_v = R_v \frac{S}{h} \quad (1)$$

Table 1. Compositions of the investigated natural rubber based composites (phr).

	NR 1	NR 2	NR 3	NR 4	NR 5	NR 6
Natural Rubber-SVR 10	100.0	100.0	100.0	100.0	100.0	100.0
Zinc Oxide	3.0	3.0	3.0	3.0	3.0	3.0
Stearic Acid	2.0	2.0	2.0	2.0	2.0	2.0
Carbon Black N 330	70.0	-	-	-	-	-
N 330/3	-	70.0	-	-	-	-
N 330/7	-	-	70.0	-	-	-
Printex XE-2B	-	-	-	70.0	-	-
Pr/Si 3	-	-	-	-	70.0	-
Pr/Si 7	-	-	-	-	-	70.0
TBBS	1.5	1.5	1.5	1.5	1.5	1.5
Sulfur	2.0	2.0	2.0	2.0	2.0	2.0

where:

R_v—ohmic resistance between the electrodes; h—sample thickness between the electrodes, m; S—cross sectional area of the measuring electrode, m^2.

A Wheatstone bridge was used to measure the resistance.

2) Surface Resistivity

The surface resistivity of the samples was calculated using Equation (2):

$$\rho_s = 100R_s \tag{2}$$

where R_s is the resistivity measured in Ω.

2.4.2. Dielectric Properties Measurement

The dielectric properties of the composite materials were measured by the resonant perturbation method. According to the resonant perturbation method, the tested sample is introduced into a resonator, and the electromagnetic parameters of the sample are deduced from the change in the resonant frequency and quality factor of the resonator [26] . For the rectangular cavity, TE_{10n}modes are used for the complex permittivity measurements. The sample was placed on the spot of maximum intensity of electric field, where n = odd is always adopted, because the sample's position can be located easily as the geometric centre of the cavity is one of the maximum positions [27] . The formulas for the real and imaginary parts of the relative permittivity are as follows [28] :

$$\varepsilon' = \left(\frac{f_c - f_s}{2f_s}\right) \cdot \left(\frac{V_c}{V_s}\right) + 1 \tag{3}$$

$$\varepsilon'' = \left(\frac{V_c}{4V_s}\right) \cdot \left(\frac{1}{Q_s} - \frac{1}{Q_c}\right) \tag{4}$$

where f_c and Q_c are resonance frequency and Q-factor of the cavity without an inserted sample, $f_s$and Q_s, are with an inserted sample, respectively; V_c is the volume of the cavity; V_s is the volume of the sample.

2.4.3. Microwave Properties Measurement

The total shielding effectiveness was defined as the ratio between the incident power on the sample P_I and the transmitted power P_T in accordance with Equation (5) [29] - [31] :

$$SE_T = 10\log\frac{P_I}{P_T} \tag{5}$$

The total shielding effectiveness (SE_T, in dB) and the reflective shielding effectiveness of the sample surface (SE_R, in dB) were determined by Equations (6) and (7) [32] - [35] :

$$SE_T = -10\lg T, \tag{6}$$

where $T = |P_T/P_I| = |S_{21}|^2$.

$$SE_R = -10\lg(1-R), \tag{7}$$

where $R = |P_R/P_I| = |S_{11}|^2$.

S_{11} and S_{21} are the complex scattering parameters or S-parameters (S_{11} corresponds to the reflection coefficient and S_{21}-to the transmission coefficient).

The absorptive shielding effectiveness (SE_A) was calculated as the difference between (6) and (7), as shown in Equation (8):

$$SE_A = SE_T - SE_R \tag{8}$$

The attenuation coefficient (α, dB/cm) was determined using Equation (8). Substituting SE_T and SE_R in it with Equations (6) and (7), Equation (9) is obtained.

$$SE_A = 10\lg\left(\frac{P_I\left(1-|S_{11}|^2\right)}{P_T}\right) \tag{9}$$

For determining the attenuation coefficient, Equation (9) was divided by the thickness of the sample d in centimetres, as shown in Equation (10).

$$\alpha = \frac{SE_A}{d} = 10\lg\left(\frac{P_T}{P_I\left(1-|S_{11}|^2\right)}\right) \tag{10}$$

The reflection losses were determined according to Equation (11).

$$R_L = 20\log_{10}|\Gamma| \tag{11}$$

where R_L is the reflection loss, $|\Gamma|$ is the reflection coefficient.

The measurements were carried out at room temperature varying from 19°C up to 24°C and incident power P_I at the inlet of the coaxial measuring line varying from 800 µW up to 1300 µW within the frequency range of 1 GHz to 12 GHz.

3. RESULTS AND DISCUSSION

3.1. Electrical Properties

Most elastomers are dielectrics, i.e. they have low conductivity (less than 10^{-11} S/m). The introduction of various conductive nanofillers might improve considerably their conductivity due to the percolation pathways in the elastomer matrix built by the conductive filler. Thus the electroconductivity increases drastically at a certain filler concentration termed as "percolation threshold". That widens to a great extent the application areas of elastomer composites [36] - [39] .

Figure 1 presents the filler concentration dependence of volume resistivity and surface resistivity. Our previous studies have shown 7.5 phr to be the critical concentration of Printex XE-2B in the rubber matrix at which the percolation threshold is reached [40] . However, in the case aiming at determining the effect that the second-dielectric phase (silica) in the studied hybrid fillers has, the filler concentration in the studied composites is much higher than the one needed to pass the percolation threshold. As Figure 1 shows, ρ_v and ρ_s values for the composites comprising carbon black N 330 (NR 1) and for those comprising hybrid fillers on its base (NR 2 and NR 3) vary in the range 2.2×10^3 to 3.8×10^3 Ω·m and 2.3×10^6 - 6.0×10^6 Ω, respectively. The composites comprising conductive carbon black Printex XE-2B (NR 4) and those comprising hybrid fillers on its base (NR 5 and NR 6) have much lower ρ_v and ρ_s values—3.6 to 4.9 Ω·m and 5.2×10^2 - 6.3×10^2Ω, respectively, i.e. those composites have higher conductivity and can be classified as true conductors. The well pronounced difference in ρ_v and ρ_s values for the composites is due to the difference in the size of carbon black particles, to the difference in the their specific surface area, respectively. The specific surface area (BET) of furnace carbon black N 330 is 75 m^2/g, that of Printex XE-2B-1000 m^2/g [40] . The smaller particles of Printex XE-2B, their higher specific surface area, respectively, favours the formation of conductive pathways, even at lower filler concentrations what is a prerequisite for higher conductivity.

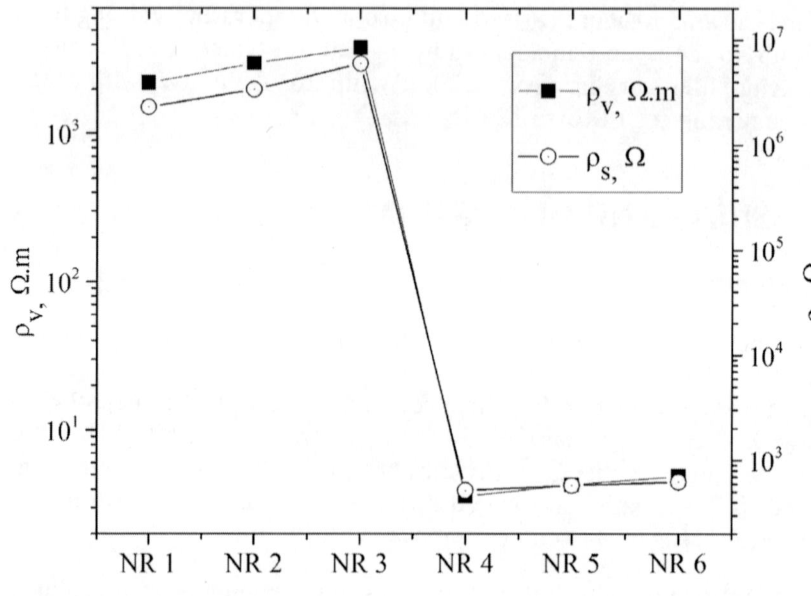

Figure 1. Volume resistivity and (ρ_v) surface resistivity (ρ_s) of the studied composites.

Evidently in all cases, the volume resistivity and surface resistivity values for the composites comprising the investigated hybrid fillers increase with the increasing silica amount in them, i.e. their conductivity decreases. That is due to the silica phase penetration amongst the chain-like structures of carbon black [41] . Thus, the dielectric phase insulates the conductive one and the formation of conductive pathways is hindered, hence conductivity of the composite is lower. The results obtained are in full accordance with our working hypothesis.

3.2. Dielectric Properties

Figure 2 and Figure 3 present the frequency dependencies of the real part (ε_r') and imaginary part (ε'') of permittivity of the studied composites.

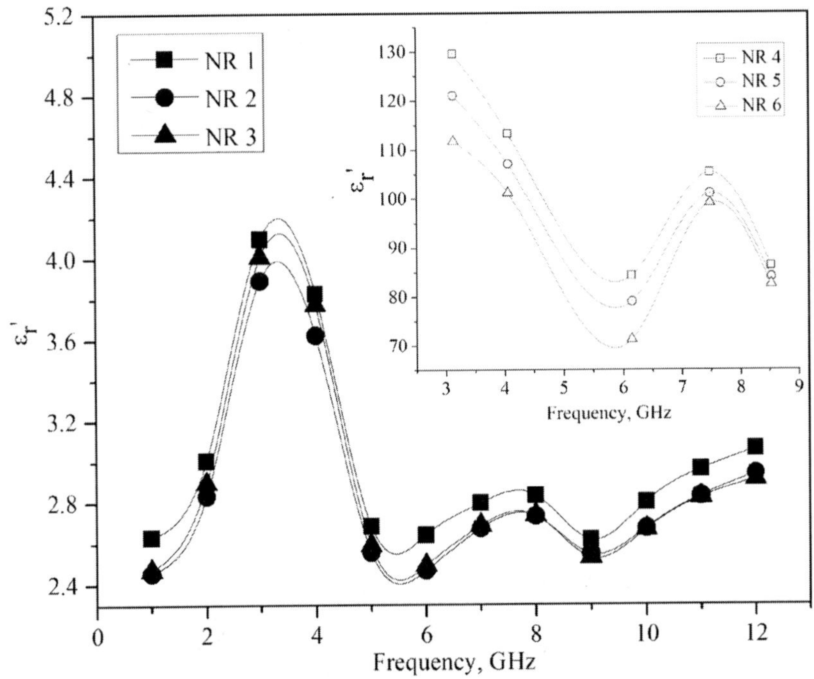

Figure 2. Frequency dependence of the real part of permittivity (ε_r') of the studied composites.

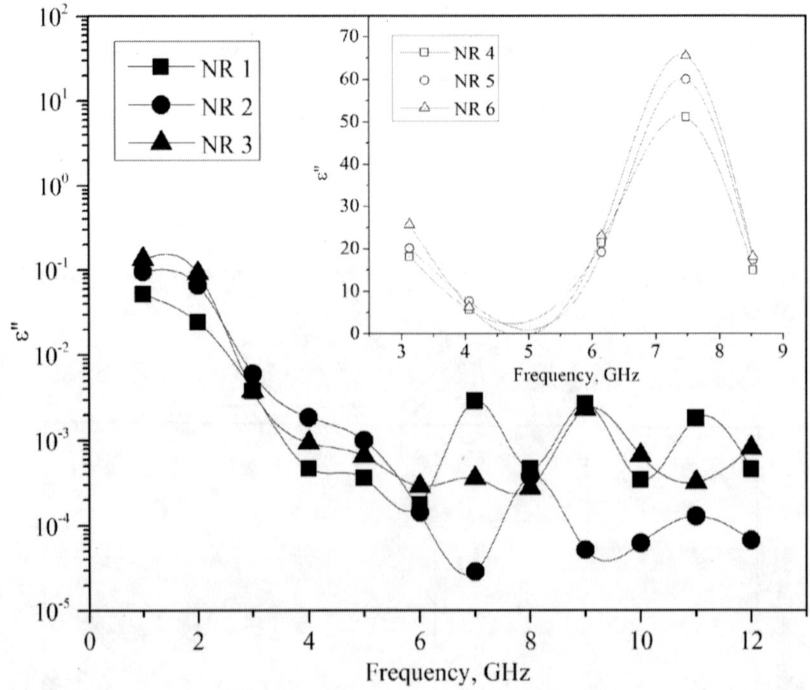

Figure 3. Frequency dependence of the imaginary part of permittivity (ε") of the studied composites.

As Figure 2 shows, in the entire frequency range, ε_r' values for NR 1 comprising virgin carbon black are the highest of those for the composites comprising furnace carbon black N 330 and the hybrid fillers on its base. That is due to the interfacial polarization resulting from the inhomogeneity introduced by the filler particles and the conductive pathways they have built. The higher ε_r' values in the case reveal that, the molecular polarization is accomplished easily [42] . The ε_r' values of composites NR 2 and NR 3 comprising the hybrid fillers prepared by impregnation of furnace carbon black N 330 with silicasol are slightly lower than those of composite NR 1 filled with the substrate N 330 carbon black. The penetration of the silica phase amongst the chain structures of carbon black and their isolation hinders the formation of conductive pathways. That hampers the molecular polarization, therefore the real part of permittivity of the composites obtained is lower [41] . The higher dielectric silica phase amount in the hybrid fillers, the more pronounced its impact is. Similar effects have been observed for the composites comprising Printex XE-2B conductive carbon black and the hybrid fillers based on it (NR 4, NR 5 and NR 6). However, in the case the dielectric constant values are much higher (ε_r' varies from 112 to 130 at 3 GHz) than those for the composites filled with furnace carbon black (ε_r' varies from 3.8 to 4.2 at 3 GHz). That is due to the difference in the particle size of the two fillers and to the difference in their ability to form chain structures (conductive pathways). The smaller size of Printex XE-2B carbon black particles and their aptitude to form chain structures favour the formation of conductive pathways. Hence, the interfacial polarization resulting from inhomogeneity introduced by the filler particles becomes easier and consequently the real part of permittivity gets higher.

Figure 3 shows ε" to be a parameter very sensitive to frequency. As seen, ε'' values for the composites filled with conductive carbon black Printex XE-2B are higher than those for composites with furnace carbon black N 330. The result may also be assigned to the differences in their structurality and main characteristics.

3.3. Microwave Properties

Figure 4 and Figure 5 present the frequency dependencies of the reflection loss and coefficient of attenuation of the composites studied.

As Figure 4 shows, the reflection loss values for the composites filled with furnace carbon black N 330 (NR 1), as well as for those comprising fillers on its base (NR 2 and NR 3), as determined by the amount of the second (dielectric) phase, do not differ significantly. Apparently, in the case, the low values (less than −10 dB) are observed in the 1 GHz to 3 GHz bandwidth. In the 3 GHz to 12 GHz range the reflective loss values are higher than −10 dB. That indicates matching between the impedance of free space and the impedance of the investigated material only in the 1 GHz to 3 GHz frequency range, i.e. the electromagnetic wave penetrates the sample. At higher frequencies the greater portion of the electromagnetic wave is reflected by the sample. As seen from

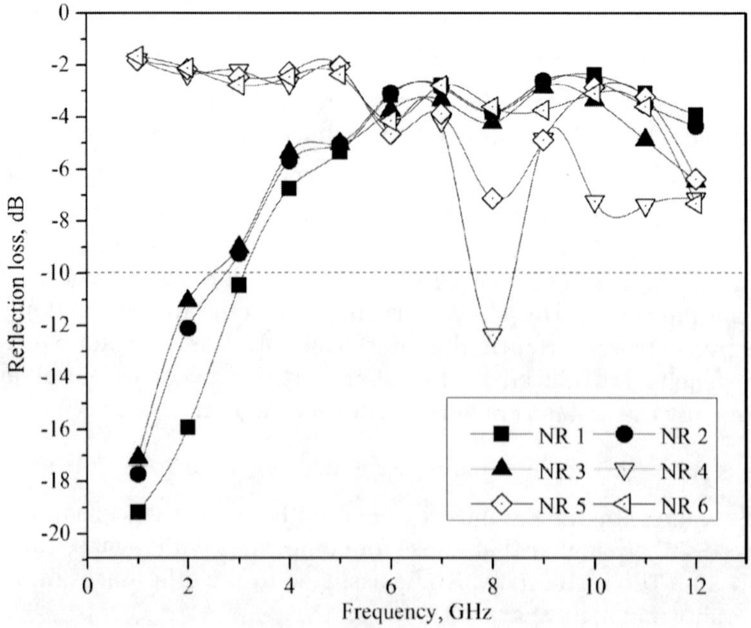

Figure 4. Reflection loss of the studied composites.

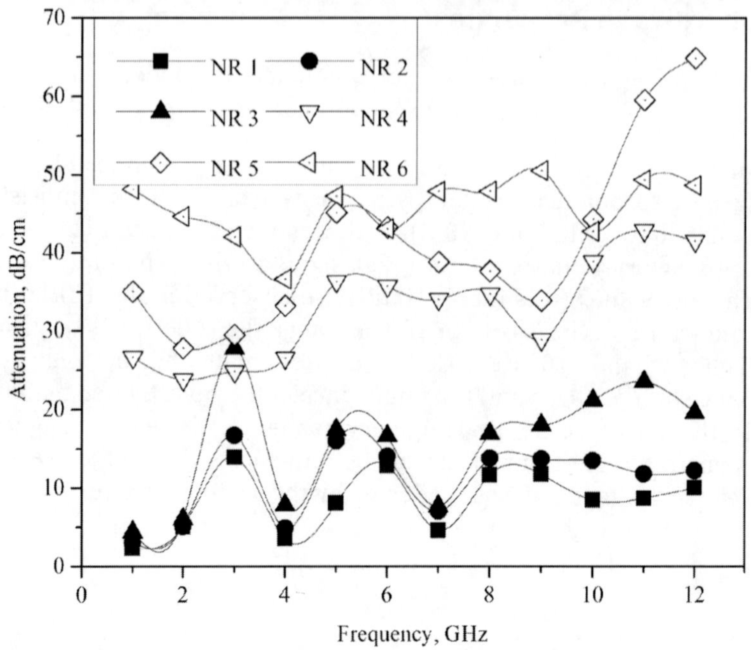

Figure 5. Attenuation coefficient of the studied composites.

Figure 5, the second phase has a better pronounced effect upon the attenuation coefficient. Evidently, the higher silica amounts improve the electromagnetic attenuation. Nonetheless, the attenuation coefficient values for the composites investigated are too low, while the ones for the reflection are too high. Hence, the composites filled with furnace carbon black N 330 and those with hybrid fillers on its base are unsuitable microwave absorbing materials. To produce good microwave absorbers one needs composites of low reflection and good attenuation properties. There are two methods for achieving low reflection. The first one involves making so that the electromagnetic waves are attenuated completely by the composite. The method is hard to accomplish, therefore researchers prefer the second one which consists of letting the electromagnetic wave penetrate the composite as much as possible and be attenuated by the absorbing material. Practically that can be realized by matching the impedance of free space with the material's impedance. The necessary condition is that the real part of magnetic permeability (μ') and the real part of the dielectric permittivity (ε_r') should be equal, i.e. $\mu'/\varepsilon_r' = 1$ [12] , [43] . Keeping in mind that, carbon black does not possess magnetic properties, the condition cannot be fulfilled in the case studied because the dielectric constant of the composites comprising furnace carbon black N 330 and the hybrid fillers based on it is much higher than their magnetic permeability [44] . When matching between the impedance of free space and material's impedance is achieved, i.e. the electromagnetic wave having penetrated the material should be attenuated, the necessary condition is that, the imaginary part of permittivity (ε'') and the imaginary part of permeability (μ'') should be higher [12] , [43] .

As Figure 4 shows, the reflection loss values for the composites comprising conductive carbon black Printex XE-2B and the hybrid fillers on its base in all cases are higher than −10 dB. In the case that means the impedance of the free space mismatches the impedance of the material, i.e. the samples reflect the greater portions of incident electromagnetic waves. The fact is due to the higher conductivity of the fillers studied which is closer to the one metals have. However, as seen fromFigure 5, the composites comprising conductive carbon black Printex XE-2B and the hybrid fillers on its base have electromagnetic attenuation values much higher than those for the composites filled with furnace carbon black N 330. The tendency of an increasing attenuation coefficient with the increasing silica amount in the hybrid fillers has also been observed in the case. That is due to the penetration of the silica phase amongst the chain structures of carbon black which isolation hinders the formation of conductive pathways.

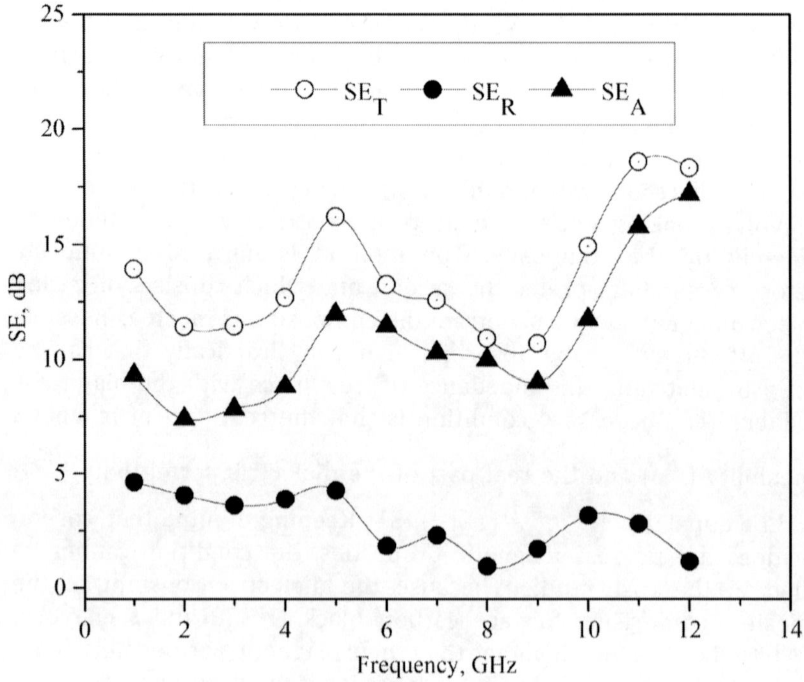

Figure 6. Frequency dependence of electromagnetic interference shielding effectiveness of the composite NR 5.

The more pronounced ability of the composites comprising conductive carbon black Printex XE-2B and the hybrid fillers on its base to attenuate the electromagnetic waves could be explained examining thoroughly the reflection losses presented in Figure 4 and the electromagnetic interference shielding effectiveness presented in Figure 6. As seen from Figure 4, the reflection loss for the composite comprising hybrid filler based on Printex XE-2B and 3% of silica (NR 5), for instance, at 1 GHz is about −1.8 dB, i.e. the value of the reflection coefficient |Γ| is about 0.81. The amplitude of the reflected wave will be approximately 81% of the incident wave. That means the power of the reflected wave will be 66% of the power of the incident wave (P_I). According to Equation (7) and Figure 6, the reflective shielding effectiveness in that case is about 4.6 dB at 1 GHz. According to the law of conservation of energy, the wave penetrating the sample will have power equal to 34% of that of the incident wave. In accordance with Figure 6, the total shielding effectiveness (SE_T) at 1 GHz is approximately 14 dB. As it is clear from Equation (6), the transmitted power (P_T) is approximately 4% of the incident power (P_I). According to Equations (6), (8) and (10), the low P_T value determines the high SE_A value, the higher attenuation, as well as the dominating absorption character of the microwave shielding effectiveness. The more pronounced ability of the composites comprising conductive carbon black Printex XE-2B and the hybrid fillers on its base to attenuate the electromagnetic waves could be also explained by their higher values of the imaginary part of permittivity. Thus they meet the requirements for a good attenuation of electromagnetic waves. Though, there is a noteworthy drawback in the case—the undesired greater dielectric heating resulting from the higher ε'' values [43] .

Figure 7 presents the frequency dependence of electromagnetic interference shielding effectiveness of the composite, comprising the hybrid filler based on furnace carbon black N 330 and 3% of silica. As seen, the total shielding effectiveness values of the composites filled with furnace carbon black are about three times lower than those of the composites comprising conductive carbon black. It could be said in the case that, reflection is the predominating component in the total shielding effectiveness.

The composites obtained, those comprising conductive carbon black Printex XE-2B and the hybrid fillers on its base, in particular, could find wide application in various areas of electronics. Their total shielding effectiveness values (varying between 14 and 19 dB), guarantee electromagnetic shielding of 95% and make them valuable for manufacturing electromagnetic interference shielding materials [45] - [47] . Their high electric conductivity and relatively high ability to attenuate the electromagnetic waves, make them suitable for other applications as well—namely as antennas for mobile devices, automobiles, sensors and military-related applications. Of course, the applications could be much more provided that, the properties of the composites discussed above are improved. It is affordable by a more precise optimization of the amount ratio between the conductive and dielectric phases in the hybrid fillers, as well as by appropriating concentration of those fillers in the elastomer composites.

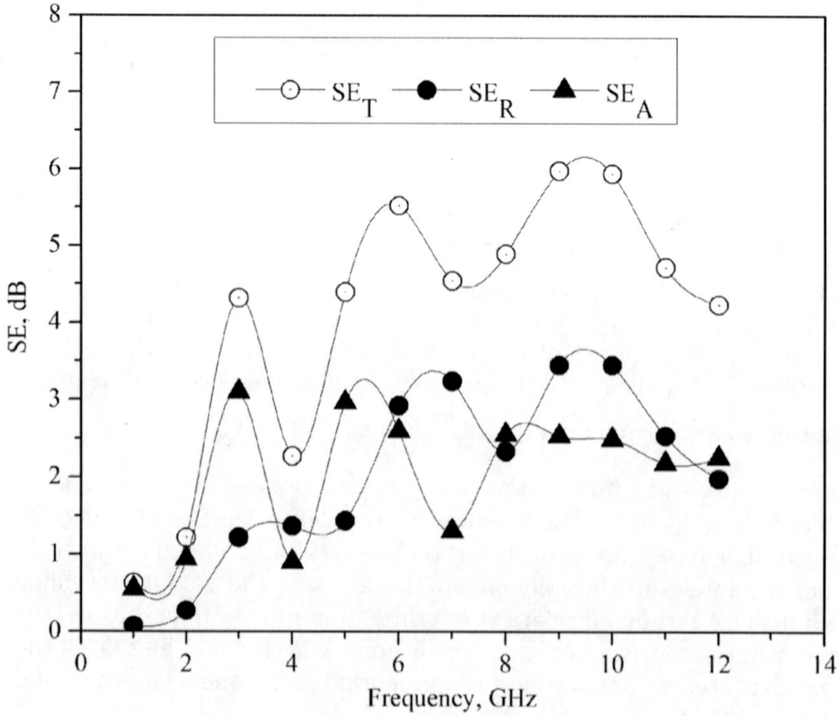

Figure 7. Frequency dependence of electromagnetic interference shielding effectiveness of the composite NR 2.

4. CONCLUSION

A comparison has been made of the electric, dielectric and microwave properties of natural rubber based composites comprising dual phase fillers based on furnace carbon black or conductive carbon black and different amounts of silica. It has been established that, the specifics of the carbon phase have a marked strong effect upon the properties mentioned above. The interpenetration of the two filler phases and the grade of isolation of the conductive carbon phase by the dielectric one depend on the ratio between them. On the other hand, that leads to a change in all properties of the studied composites, what allows tailoring those characteristics. With an increasing amount of the dielectric phase in the hybrid fillers, the electric conductivity and dielectric constant of the composites get lower, while their ability to absorb electromagnetic waves enhances. The higher electroconductivity and the relatively high ability of the composites comprising conductive carbon black to attenuate the electromagnetic waves, make them suitable for several applications such as electromagnetic interference shielding materials, antennas for mobile devices, automobiles, sensors and military-related applications, as they meet the requirements set for the properties that such type of materials should possess.

ACKNOWLEDGEMENTS

The work is a part of a project, funded by King Abdulaziz University, Saudi Arabia under grant number MB/11/12/436. The authors acknowledge the technical and financial support.

REFERENCES

1. Ghasemi, A., Liu, X. and Morisako, A. (2007) Magnetic and Microwave Absorption Properties of BaFe12-x (Mn0.5Cu0.5Zr)x/2O19 Synthesized by Sol-Gel Processing. Journal of Magnetism and Magnetic Materials, 316, e105-e108.

2. Yusoff, A.N., Abdullah, M.H., Ahmad, S.H., Jusoh, S.F., Mansor, A.A. and Hamid, S.A.A. (2002) Electromagnetic and Absorption Properties of Some Microwave Absorbers. Journal of Applied Physics, 92, 876-882.

3. Chung, D.D.L. (2000) Materials for Electromagnetic Interference Shielding. Journal of Materials Engineering and Performance, 9, 350-354.

4. Chung, D.D.L. (2012) Carbon Materials for Structural Self-Sensing, Electromagnetic Shielding and Thermal Interfacing. Carbon, 50, 3342-3353.

5. Qin, F. and Brosseau, C. (2012) A Review and Analysis of Microwave Absorption in Polymer Composites Filled with Carbonaceous Particles. Journal of Applied Physics, 111, Article ID: 061301.

6. Sunny, V., Kurian, P., Mohanan, P., Joy, P.A. and Anantharaman, M.R. (2010) A Flexible Microwave Absorber Based on Nickel Ferrite Nanocomposite. Journal of Alloys and Compounds, 489, 297-303.

7. Jamal, E.M.A., Mohanan, P., Joy, P.A., Kurian, P. and Anantharaman, M.R. (2009) Effect of Nickel Nanofillers on the

Dielectric and Magnetic Properties of Composites Based on Rubber in the X-Band. Applied Physics A, 97, 157-165.

8. Prema, K.H., Kurian, P., Anantharaman, M.R., Suma, M.N. and Joseph, M. (2008) Permittivity Characteristics in the X- and S-Band Frequencies of Microwave Absorbers Based on Rubber Ferrite Composites. Journal of Elastomers and Plastics, 40, 331-346.

9. Sunny, V., Kumar, D.S., Mohanan, P. and Anantharaman, M.R. (2010) Nickel/Carbon Hybrid Nanostructures as Microwave Absorbers. Materials Letters, 64, 1130-1132.

10. Miyata, Y. and Matsumoto, M. (1997) Two-Layer Wave Absorber Composed of Soft-Magnetic and Ferroelectric Substances. IEEE Transactions on Magnetics, 33, 3427-3429.

11. Feng, Y., Qiu, T., Li, X. and Shen, C. (2007) Microwave Absorption Properties of the Carbonyl Iron/EPDM Radar Absorbing Materials. Journal of Wuhan University of Technology-Mater. Sci. Ed., 22, 266-270.

12. Duggal, S. and Aul, G.D. (2014) Review on Effect of Electric Permittivity and Magnetic Permeability Over Microwave Absorbing Materials at Low Frequencies. International Journal of Engineering and Advanced Technology (IJEAT), 3, 12-19.

13. Barba, A.A., Lamberti, G., d'Amore, M. and Acierno, D. (2006) Carbon Black/Silicone Rubber Blends as Absorbing Materials to Reduce Electro Magnetic Interferences (EMI). Polymer Bulletin, 57, 587-593.

14. Joshi, A. and Datar, S. (2015) Carbon Nanostructure Composite for Electromagnetic Interference Shielding. Pramana, 84, 1099-1116.

15. Krishnan, Y., Chandran, S., Usman, N., Smitha, T.R., Parameswaran, P.S. and Prema, K.H. (2015) Processability, Mechanical and Magnetic Studies on Natural Rubber-Ferrite Composites. International Journal of Chemical Studies, 3, 15-22.

16. Ismail, H., Sam, S.T., Mohd Noor, A.F. and Bakar, A.A. (2007) Properties of Ferrite-Filled Natural Rubber Composites. Polymer-Plastics Technology and Engineering, 46, 641-650.

17. Kong, I., Hj Ahmad, S., Hj Abdullah, M., Hui, D., NazlimYusoff, A. and Puryanti, D. (2010) Magnetic and Microwave Absorbing

Properties of Magnetite-Thermoplastic Natural Rubber Nanocomposites. Journal of Magnetism and Magnetic Materials, 322, 3401-3409.

18. Urogiova, E., Hudec, I. and Bellusova, D. (2006) Magnetic and Mechanical Properties of Strontium Ferrite-Rubber Composites. Kautschuk Gummi Kunststoffe, 59, 224-228.

19. Dishovsky, N. (2009) Rubber Based Composites with Active Behaviour to Microwaves (Review). Journal of the University of Chemical Technology and Metallurgy, 44, 115-122.

20. Yuan, Y., Zhang, S.R., Zhou, X.H. and Li, E.Z. (2013) MgTiO3 Filled PTFE Composites for Microwave Substrate Applications. Materials Chemistry and Physics, 141, 175-179.

21. Fan, B.-H., Zha, J.-W., Wang, D., Zhao, J. and Dang, Z.-M. (2012) Size-Dependent Low-Frequency Dielectric Properties in the BaTiO3/Poly (Vinylidene Fluoride) Nanocomposite Films. Applied Physics Letters, 100, Article ID: 012903.

22. Jin, S., Wang, L., Wang, Z., Huang, B., Zhang, Q. and Fu, Z. (2015) Dielectric Properties of Modified SrTiO3/PTFE Composites for Microwave RF Antenna Applications. Journal of Materials Science: Materials in Electronics, 26, 7431-7437.

23. Ullah, U., Ali, W.F.F.W., Ain, M.F., Mahyuddin, N.M. and Ahmad, Z.A. (2015) Design of a Novel Dielectric Resonator Antenna Using MgTiO3-CoTiO3 for Wideband Applications. Materials & Design, 85, 396-403.

24. Wang, M.J., Kutsovsky, Y., Zhang, P., Mehos, G., Murphy, L.J. and Mahmud, K. (2002) Using Carbon-Silica Dual Phase Filler Improve Global Compromise between Rolling Resistance, Wear Resistance and Wet Skid Resistance for Tires. Kautschuk Gummi Kunststoffe, 55, 33-40.

25. Rattanasom, N., Saowapark, T. and Deeprasertkul, C. (2007) Reinforcement of Natural Rubber with Silica/Carbon Black Hybrid Filler. Polymer Testing, 26, 369-377.

26. Chen, L.F., Ong, C.K., Neo, C.P., Varadan, V.V. and Varadan, V.K. (2004) Microwave Electronics: Measurement and Materials Characterization. John Wiley & Sons Ltd., Chichester.

27. Sheen, J. (2009) Measurements of Microwave Dielectric Properties by an Amended Cavity Perturbation Technique. Measurement, 42, 57-61.

28. Verma, A. and Dube, D.C. (2005) Measurement of Dielectric Parameters of Small Samples at X-Band Frequencies by Cavity Perturbation Technique. IEEE Transactions on Instrumentation and Measurement, 54, 2120-2123.

29. Jana, P.B., Mallick, K. and De, S.K. (1992) Effects of Sample Thickness and Fiber Aspect Ratio on EMI Shielding Effectiveness of Carbon Fiber Filled Polychloroprene Composites in the X-Band Frequency Range. IEEE Transactions on Electromagnetic Compatibility, 34, 478-481.

30. Paul, C.R. (2006) Introduction to Electromagnetic Compatibility. 2nd Edition, John Wiley & Sons Inc., Hoboken.

31. Ott, H.W. (2009) Electromagnetic Compatibility Engineering. 1st Edition, John Wiley & Sons Inc., Hoboken.

32. Hernandez, B. (2013) Effect of Graphitic Carbon Nanomodifiers on the Electromagnetic Shielding Effectiveness of Linear Low Density Polyethylene Nanocomposites. Ph. D. Thesis, the Graduate School of Clemson University, Clemson.

33. Srikanth, V.V.S.S. and Raju, K.C.J. (2015) Graphene/Polymer Nanocomposites as Microwave Absorbers. In: Sadasivuni, K.K., Ed., Graphene-Based Polymer Nanocomposites in Electronics, Springer International Publishing, Burlin, 307-343.

34. Sung-Hoon, P., Theilmann, P.T., Asbeck, P.M. and Bandaru, P.R. (2010) Enhanced Electromagnetic Interference Shielding through the Use of Functionalized Carbon-Nanotube-Reactive Polymer Composites. IEEE Transactions on Nanotechnology, 9, 464-469.

35. Hong, Y.K., Lee, C.Y., Jeong, C.K., Lee, D.E., Kim, K. and Joo, J. (2003) Method and Apparatus to Measure Electromagnetic Interference Shielding Efficiency and Its Shielding Characteristics in Broadband Frequency Ranges. Review of Scientific Instruments, 74, 1098-1102.

36. Rahaman, M., Chaki, T.K. and Khastgir, D. (2011) Development of High Performance EMI Shielding Material from EVA, NBR, and

Their Blends: Effect of Carbon Black Structure. Journal of Materials Science, 46, 3989-3999.

37. Sohi, N.J.S., Rahaman, M. and Khastgir, D. (2011) Dielectric Property and Electromagnetic Interference Shielding Effectiveness of Ethylene Vinyl Acetate-Based Conductive Composites: Effect of Different Type of Carbon Fillers. Polymer Composites, 32, 1148-1154.

38. Sohi, N.J.S., Bhadra, S. and Khastgir, D. (2011) The Effect of Different Carbon Fillers on the Electrical Conductivity of Ethylene Vinyl Acetate Copolymer-Based Composites and the Applicability of Different Conductivity Models. Carbon, 49, 1349-1361.

39. Selvin Thomas, P., Abdullateef, A.A., Al-Harthi, M.A., Atieh, M.A., De, S.K., Rahaman, M., Chaki, T.K., Khastgir, D. and Bandyopadhyay, S. (2011) Electrical Properties of Natural Rubber Nanocomposites: Effect of 1-Octadecanol Functionalization of Carbon Nanotubes. Journal of Materials Science, 47, 3344-3349.

40. Al-Hartomy, O.A., Al-Solamy, F., Al-Ghamdi, A., Dishovsky, N., Ivanov, M., Mihaylov, M. and El-Tantawy, F. (2011) Influence of Carbon Black Structure and Specific Surface Area on the Mechanical and Dielectric Properties of Filled Rubber Composites. International Journal of Polymer Science, 2011, Article ID: 521985.

41. Al-Ghamdi, A.A., Al-Hartomy, O.A., Al-Solamy, F.R., Dishovsky, N., Malinova, P. and Lakov, L. (2015) Characterization of Hybrid Fillers Based on Carbon Black of Different Types Obtained by Impregnation. Proceedings of the Institution of Mechanical Engineers, Part L: Journal of Materials: Design and Applications, In Press.

42. Puryanti, D., Ahmad, S.H. and Abdullah, M.H. (2006) Effect of Nickel-Cobalt-Zinc Ferrite Filler on Electrical and Mechanical Properties of Thermoplastic Natural Rubber Composites. Polymer-Plastics Technology and Engineering, 45, 561-567.

43. Zhang, B., Yong, F., Jie, X., Yang, Y. and Lu, H. (2006) Microwave-Absorbing Properties of De-Aggregated Flake-Shaped Carbonyl-Iron Particle Composites at 2 - 18 GHz. IEEE Transactions on Magnetics, 42, 1778-1781.

44. Xie, S., Yang, Y., Hou, G., Wang, J. and Ji, Z. (2016) Development of Layer Structured Wave Absorbing Mineral Wool Boards for Indoor Electromagnetic Radiation Protection. Journal of Building Engineering, 5, 79-85.

45. Faisal, M. and Khasim, S. (2014) Ku-Band EMI Shielding Effectiveness and Dielectric Properties of Polyaniline-Y2O3 Composites. Polymer Science Series A, 56, 366-372.

46. Wu, Z.P., Li, M.M., Hu, Y.Y., Li, Y.S., Wang, Z.X., Yin, Y.H., Chen, Y.S. and Zhou, X. (2011) Electromagnetic Interference Shielding of Carbon Nanotube Macrofilms. Scripta Materialia, 64, 809-812.

47. Dhawan, S.K., Singh, N. and Venkatachalam, S. (2002) Shielding Behaviour of Conducting Polymer-Coated Fabrics in X-Band, W-Band and Radio Frequency Range. Synthetic Metals, 129, 261-267.

5

STRAIN ROCKBURSTS SIMULATED BY LOW-STRENGTH BRITTLE EQUIVALENT MATERIALS

Lang Li,[1,2] Mingyang Wang,[1] Pengxian Fan,[1,2] Haiming Jiang,[1] Yihao Cheng,[1] and Derong Wang[1]

[1]State Key Laboratory for Disaster Prevention & Mitigation of Explosion & Impact, PLA University of Science and Technology, Nanjing 210007, China

[2]Equipment Academy of Rocket Army, Beijing 100085, China

ABSTRACT

This paper presents experimental study on rockbursts that occur in deep underground excavations. To begin with, the boundary conditions for excavation in deep underground engineering were analysed and elastic adaptive boundary is an effective way to minimize the boundary effect of geomechanical model test. Then, in order to simulate an elastic adaptive loading boundary, Belleville springs were used to establish this loading boundary. With the aforementioned experimental set-ups and fabrication of similarity models for test, the phenomena of strain mode rockbursts were satisfactorily reproduced in laboratory. The internal stress, strain, and convergences of the openings of the model were instrumented by subtly preembedded sensors and transducers. Test results showed that, with an initial state of high stress from both upper layers' gravitational effects and in situ stress due to tectonic movements, the excavation brings a dramatic rise in the hoop stress and sharp drop in radial stress, which leads to the splitting failure of rock mass. Finally a rockburst occurred associated with the release of strain energy stored in highly stressed rock mass. In addition, the failure of the surrounding rock demonstrated an obvious hysteresis effect which supplies valuable guide and reference for tunnel support. Not only do these results provide a basis for further comprehensive experiments, but also the data can offer assisting aids for further theoretical study of rockbursts.

1. INTRODUCTION

With the rapid development of construction and the overwhelming demands for resources, the development and use of underground spaces gradually reach into deeper areas, including mines, tunnels, and nuclear waste disposal sites. With incremental increases in the buried depth of projects, we find that some nonlinear deformations and failure phenomena occur which are different from those in shallow rock engineering. Also, they cannot be satisfactorily explained by the traditional theory of continuum mechanics. The special phenomena such as rockbursts, zonal disintegration, and anomalously low friction have sparked widespread concerns in the international rock mechanics engineering community. The study of these problems has become an important issue in the past decades [1–4]. Generally, deep underground engineering is surrounded by a geological environment with high stress, high temperature, and high water pressure which change the mechanical properties of the deep rock mass and result in phenomena such as rockbursts happening more frequently. Rockbursts occurred frequently in the process of mining and the construction of hydraulic projects. They not only pose a threat to the safety of construction workers and equipment, but also affect the construction process. Thus the research on rockbursts is imperative.

Current theories, such as energy theory [5], stiffness theory [6, 7], strength theory [8], impact tendency theory [8–10], fracture mechanics theory [11, 12], and fractal theory [13], have made some progress in the field of rockbursts. Because rockbursts happen suddenly and even randomly to some extent, these theories still cannot satisfactorily predict the occurrence of rockbursts. Therefore, a clear understanding of the exact rockburst mechanism and defining the law of rockbursts and forecasting them remain as challenging tasks.

Many experiments have been performed all over the world including uniaxial tests [7, 14, 15], biaxial loading tests [16, 17], and triaxial tests [18–25]. For example, Gu et al. [14] successfully simulated an ejective rockburst phenomenon in a uniaxial test. He et al. [21–23] studied the process of rockbursts on granite under high crustal stress. Many others combined loading tests and the loading due to human disturbance which have been conducted on rock specimens. They preliminarily come to the conclusion that dynamic loading disturbance is crucial to the occurrence of rockbursts. These tests have made great progress in the research of rockbursts, but there still exist the following weaknesses: (1) Most of the loading systems actively interfere with the failure process of the specimen. (2) In situ rockbursts, while under a three-dimensional stress state, take place because the excavation and unloading produce a free face of the opening which results in stress transformation and concentration. The above experiments produced stress strain paths that are not consistent with the stress transformation conditions of an in situ rockburst. (3) Most of the tests are based on the original rock, and the size is not big enough to satisfy the conditions. Stress concentration on the loading boundary usually leads to differences from the practical engineering. Owing to the problems listed above, laboratory studies on rockbursts have not achieved a breakthrough. What is more, many scholars have carried out research using the similar test model [26–29]. Zhou et al. [26] studied on mechanisms of slab buckling rockburst in deep tunnel using physical simulation experiment. Lu et al. [27] investigated rockburst induced by explosion stress wave at the chamber after excavation. Model test of rockburst has become an important means to study the mechanism of rockburst. However, they have not done a good job of reproducing the phenomenon of strain rockburst in the laboratory or they cannot give a complete change law for the internal stress and strain of excavation.

This paper reproduces a strain rockburst phenomenon using tests of similarity models. First, to simulate the initial stress state in deep rock mass more accurately, we suggest an elastic adaptive boundary and Belleville springs were used to establish this loading boundary. Second, by accurately simulating the process of excavating the tunnel, a free surface caused by excavation is produced, and the stress is converted due to the unloading of surrounding rock. Thus a more realistic simulation of the generation process of in situ rockbursts is carried out. The aim is to reproduce the rockburst phenomenon using tests of similarity models. Then a variety of means are used to monitor the whole occurrence and the development process of rockbursts. This can provide a basis for an experiment and data to support the further study of the initiation mechanism of rockbursts.

2. NEW LOADING BOUNDARY OF THE TEST MODEL

2.1. Boundary Conditions

A finite size model to simulate the excavation problem in an infinite medium is generally used in a geotechnical test model. Boundary conditions are essential for the success of the test. The excavation of the tunnel in practical engineering can be seen as creating a new opening with the radius in an infinite medium. It can be simplified as a plane strain problem. By taking a nearby area in rock mass as the research object, the rock mass is in direct contact with the tunnel with an area : usually reaching 6~8 according to Liao et al. [30]. The materials outside this area () can be seen as a continuous elastic medium. Its impact on the near field can be simulated by elastic elements. As a result, when creating a new opening the boundary can be equivalent to a spring model (Figure 1).

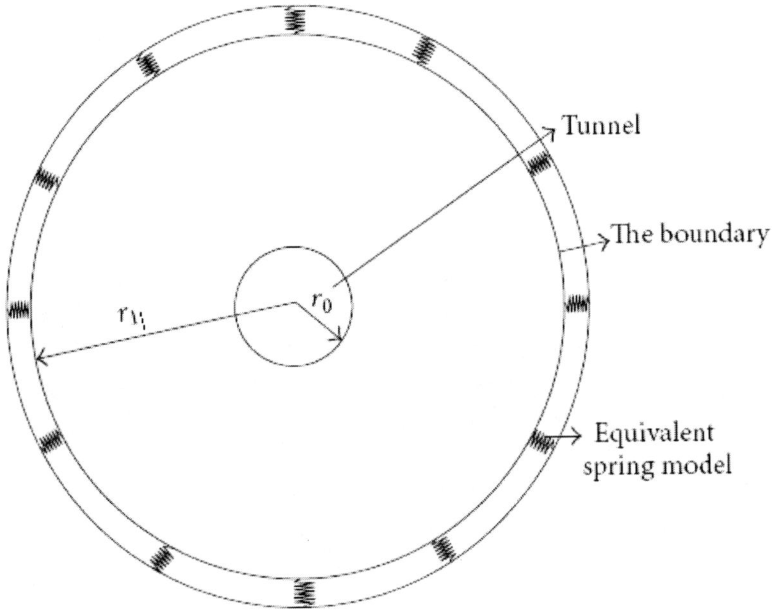

Figure 1. Equivalent spring model on the boundary of tunnel excavation.

The current boundary conditions can be classified into two kinds: one is the displacement boundary condition; the other is the stress boundary condition. The former boundary is rigid, and its displacement is even, but the stress is not uniform. For homogeneous materials it can provide satisfactory test results, but for discontinuous and heterogeneous material like a rock mass it can provide nonuniform and uncertain stress on the boundary, and thus the initial stress state of rock mass cannot be easily simulated. To ensure a uniform stress field and a more accurate simulation test boundary, the other boundary is of greater significance. As the buried depth of underground engineering increases, the stress becomes high enough so that the effect of boundary conditions on test results is very significant. In addition, deep rock masses have energetic properties [31]. As a result, these two boundary conditions cannot guarantee strict similarity between the test model and the practical engineering. So it is necessary to explore the boundary conditions to further improve the credibility of the test model in deep underground engineering. An elastic adaptive boundary is presented by Haiming et al. [32] to minimize the boundary effect of geomechanical model test. Hence, when creating a new opening in the model shown in Figure 1, the rebound stiffness on the outer boundary with radius can be calculated from

$$k = \frac{2G}{r} = \frac{E}{(1+\nu)r}, \tag{1}$$

where is Poisson's ratio, is the modulus of elasticity in shear, is elastic modulus. Therefore, during the period of the test model, the equivalent stiffness acting on the model's boundary by the loading devices should be equal to the rebound stiffness of the model due to excavation and unloading. This boundary condition is an effective way to minimize the boundary effect of geomechanical model test and is more consistent with the practical engineering.

2.2. Physical Implementation of Elastic Adaptive Boundary

A test apparatus for deep rock mass loading and unloading is used to simulate the initial in situ stress (Figure2). The vertical direction and the left- and right-hand sides of the model are uniformly loaded by the hydraulic jack. A hydraulic jack is located at the excavation direction. The bottom and front face of the model are loaded by the reaction frame loading system. Because the parameters of equivalent materials are usually very low, the rebound stiffness on the boundary of model, according to the theory, is relatively less than the test device. Therefore, the stiffness of loading system must be reduced. Belleville springs are introduced to the loading system which are set between the cylinder and the loading plate. A schematic of this arrangement is presented in Figure 3. The total pressure is provided by a hydraulic system. The pressure is transferred to the model through compressed springs and elastic energy is stored in spring systems. In this case, the elastic boundary condition can be easily satisfied.

Figure 2: Deep rock mass loading and unloading test apparatus.

According to the theory, the equivalent stiffness of the loading device acting on the boundary should be equal to the rebound stiffness of the model. By considering the bearing capacity of the Belleville spring and stiffness requirements, we select the specifications of the Belleville spring to be mm. At the same time, the deformation range of the spring in the experiment is controlled within of its maximum compression stroke for safety. According to GB/T1972–2005 [33], the specific parameters of spring are shown in Table 1. And the stiffness value of a single 250 mm spring is 6.26×10^7 N/m. The rebound stiffness (per unit area) on the border of model when excavating is 8.9×10^8 N/m^3. The composite combination of Belleville springs is adopted. It consists of four and five superimposed springs and then superimposed on the two groups together according to the relative direction. The stiffness of the Belleville spring system and its equivalent stiffness acting on the model are 1.55×10^8 N/m and 8.81×10^8 N/m^3, respectively, which can easily meet the theory [34].

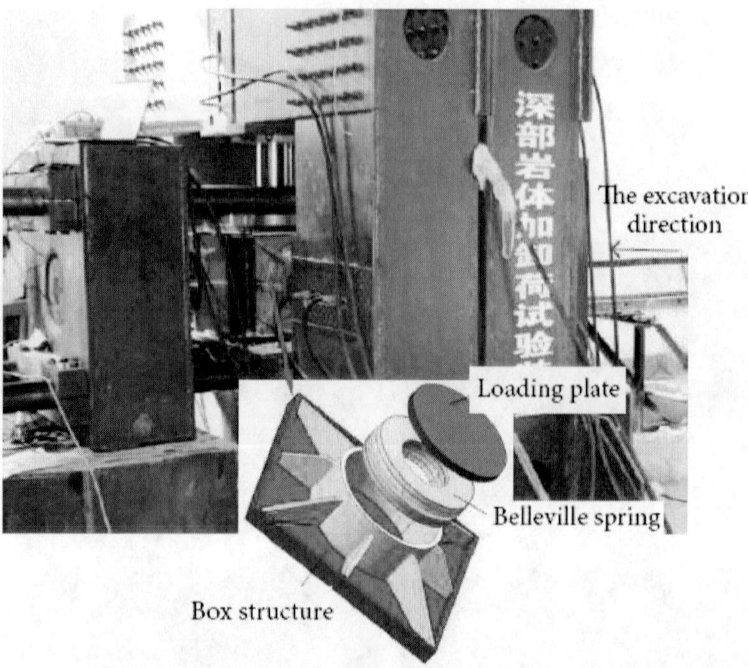

Figure 3: Schematic diagram of Belleville spring loading system.

Table 1: Belleville spring parameters.

D (mm)	d (mm)	t (mm)	h_0 (mm)	H_0 (mm)	p (KN)	f (mm)
250	127	14	5.6	19.6	249	4.2

To summarize, the test apparatus provides the following conditions at the boundary of the model: (1) the boundary can quickly supply the model with a rebound energy caused by excavation and provide the model with a proper stiffness; (2) stress can be adjusted automatically with the stiffness on the boundary.

3. TEST DESIGN AND PROCEDURES

3.1. Engineering Background

The transport tunnels of the Jinping II hydropower project are selected as the engineering prototype. This engineering project was located in the sloped terrain zone from the Tibetan Plateau to the Sichuan Basin of China. The two transport tunnels are about 17.5 km long and the cross-sectional sizes of the A and B tunnels are 5.5 × 4.5 m^2 and 6 × 5 m^2, respectively, with a line spacing of 35 m. Most of the tunnel's buried depth is more than 1500 m. The lithology of the nature rock is mostly Triassic, Limestone, and maximum tunnel depth is 2375 m. Rockbursts did occur many times during construction.

3.2. Test Scheme Design

Strain rockburst usually occurs in intact rock and is caused by deformation and failure of the rock mass [35]. In this study the test does not consider faults, joints, and cracks in the rock mass. The rock is assumed to be intact. The model dimensions (length × width × height) are 0.88 m × 0.4 m × 0.4 m. To facilitate data analysis and summarize the formulas, the opening tunnel is defined as round with a diameter of 80 mm and excavation depth of 700 mm. The measuring points for stress, strain, and temperature sensors are shown in Figure 4.

3.3. Equivalent Materials

At present the similarity criterion generally uses the elastic theory and dimensional analysis method. Specifically, the geometric similarity scale and bulk density similarity scale of the experimental model are 75 and 1, respectively. Scales of other parameters can be deduced by analogy to the equilibrium; geometric and physical equations, boundary conditions, and Newton's second law are expressed as:

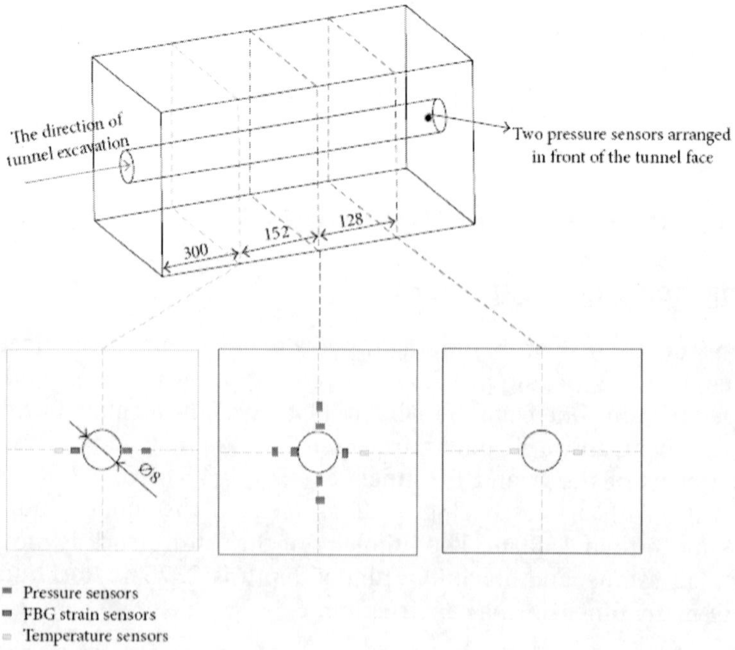

Figure 4: Stress, strain, and temperature sensors measuring points.

$$C_\sigma = C_E = C_\delta = 75,$$
$$C_t = \sqrt{75},$$
$$C_\varepsilon = C_f = C_\varphi = C_\mu = 1, \quad (2)$$

where , $C_\sigma, C_l, C_\gamma, C_E, C_\varepsilon, C_\delta, C_t, C_f, C_\varphi,$ and C_μ denote the similarity scales of stress, geometry, density, elastic modulus, strain, displacement, time, friction coefficient, friction angle, and Poisson's ratio, respectively.

Rockbursts generally occur in the hard and brittle rock. To simulate marble according to the similarity criteria, we adopt resin-based equivalent materials that are developed by Fan et al. [36]. This material uses resin as the bonding agent, barite ore and quartz sand as the aggregate, and barite powder as the fine aggregate. It is a kind of brittle materials with low-strength. The mould, specimens, curing of sample, and the mechanical parameters test such as the uniaxial compression test and shear test are shown in Figure 5. The proportion of each component used in the experiment is shown in Table 2 and the unit of each component ratio is weight percentage. The physical and mechanical parameters of the natural rock and equivalent material are shown in Table 3. It can be seen that the materials meet the theoretical requirements and do a good job of simulating the marble of the engineering prototype.

Table 2. The ratio of equivalent materials.

Components	Barite sand	Quartz sand	Barite powder	Rosin	Alcohol
Ratio	45%	5%	50%	0.5%	4.5%

Table 3. Physical and mechanical parameters of the prototype and model.

Material types	Dry density KN/m3	Elastic modulus/MPa	Compressive strength/MPa	Tensile strength/MPa	Cohesion/MPa	Internal friction angle
Marble	27.7	25300	89.2	4.95	4.24	53.1°
Equivalent materials	27.7	369.6	1.1	0.07	0.06	50.7°

3.4. Preparation of the Sample Model

The model uses the method of layered filling and compaction. To ensure a uniform stress field of the model at the boundary, flexible rubber is positioned at the contact areas between the model and test device. To decrease the friction effect, a layer of PTFE film is pasted on the other side of the flexible rubber (Figure 6(a)). The equivalent material is compressed to a dense state at 2 MPa pressure by two hydraulic jacks in the vertical direction. To avoid the formation of an obvious interface and make an effective bond the model with the material filled the next layer, and alcohol solution is sprayed on the surface of the model. Its surface is shaved with a steel rake after the softening of the material model. Then the next layer of materials is joined until the entire model is completed. The internal pressure in the model is measured by DZ-I pressure sensors, with the following dimensions: thickness 7 mm; diameter 17 mm (Figure 6(b)). The internal strain measurement used

Figure 5. Equivalent materials and mechanical experiments: (a) the mould; (b) sample of equivalent materials; (c) sample preparation; (d) maintenance of the sample; (e) uniaxial compressive test of sample; (f) shearing test of the sample.

FCC-Y FBG sensors (Figure 6(c)). First, FBG sensors are encapsulated in the equivalent material blocks and then embedded in the model after the calibration test. Because the FBG sensors are particularly sensitive to temperature, four temperature sensors are placed in the model to monitor the temperature during the process of excavation. The temperature changes very little during the test. The temperature in model ranges between 13.8 and 14.7°C. In the process of excavation, the temperature is maintained between 14.4 and 14.5°C. For the FBG strain sensor, the strain varies by roughly 10 με in the sensor for each degree change in temperature. Therefore, the variation of FBG strain sensor's wavelength is basically due to the changing stress caused by the model.

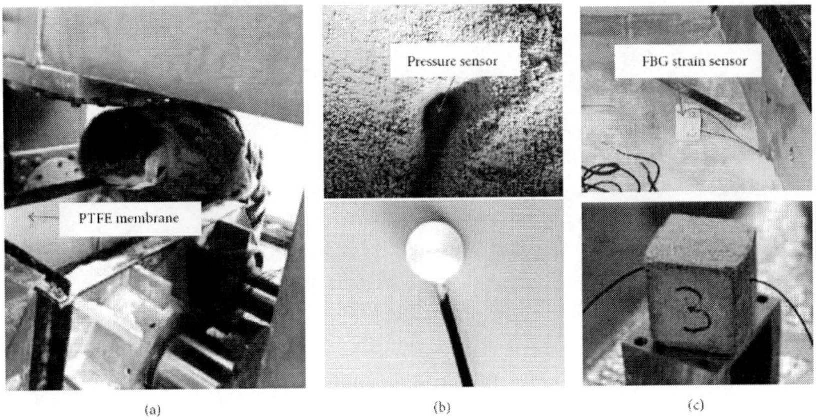

Figure 6: Fiber grating strain (FBG) sensor and pressure sensor embedded process: (a) layered filling of the model; (b) DZ-I pressure sensors; (c) FCC-Y FBG strain sensors.

3.5. Simulation of the In Situ Stress and Excavation

We set the underground engineering construction to have a buried depth of 2375 m as an example. The vertical in situ stress is given by

$$\sigma = \gamma H = 64.4\,\text{MPa}. \tag{3}$$

To simulate the in situ stress, a pressure of 0.86 MPa needs to be applied to the model. The excavated tunnel is 70 cm long and 8 cm in diameter. According to the similarity criterion, each excavation length is 7 cm and it takes 5 minutes to finish the excavation. The next step of excavation is started 15 minutes later. The spiral drill pipe driven by the motor that simulates the excavation and unloading is adopted to create a new opening in the model. The internal stress, strain, and temperature are monitored during excavation process. Excavation equipment and internal stress and strain monitoring system are shown in Figure 7. After the excavation finished, the load acting on the model's boundary is maintained for three days. At the same time, stress, strain, temperature, and tunnel convergence are monitored. The residual deformation and failure models are recorded after the dismantling of the test apparatus.

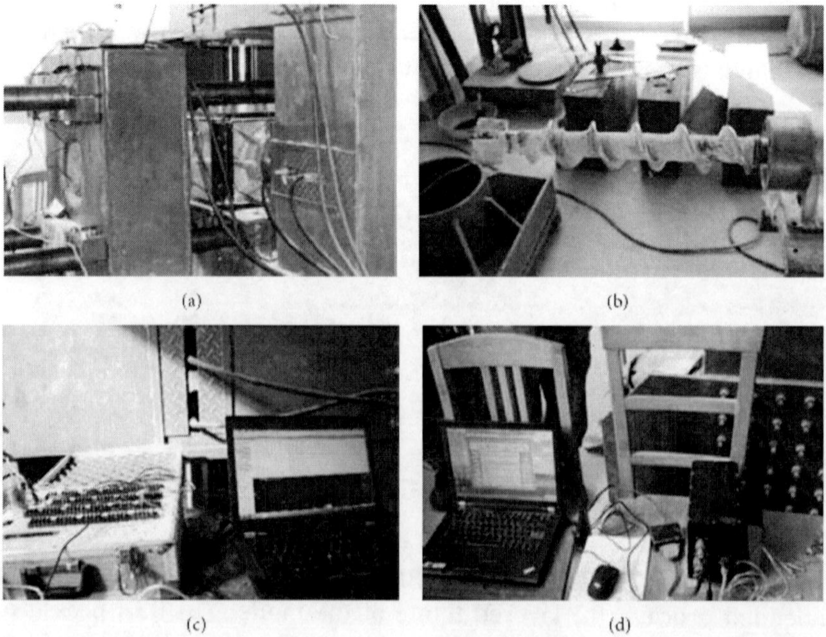

Figure 7. The process of test: (a) the process of loading; (b) the drill and excavation process; (c) internal pressure monitoring system; (d) FBG strain sensor monitoring system.

4. TEST RESULTS AND DISCUSSION

4.1. Internal Pressure

The problem of an excavated circular tunnel can be simplified as a plane strain problem. The radial pressure points, which are 2 and 8 cm from the wall as calculated by elastic theory, reduce to and , respectively. The circumferential pressure points, which are 2 and 8 cm from the wall, increase to and , respectively. Figures 8–10 show the changes of the pressure curve during the process of excavation. According to the curve, the initial pressure in each point is close to the initial in situ stress 0.86 MPa, which roughly satisfied the requirement of simulating the initial in situ stress environment for practical engineering. The pressure changing law of the surrounding rock during the excavation is similar to that calculated by elastic theory but also displays some new phenomena.

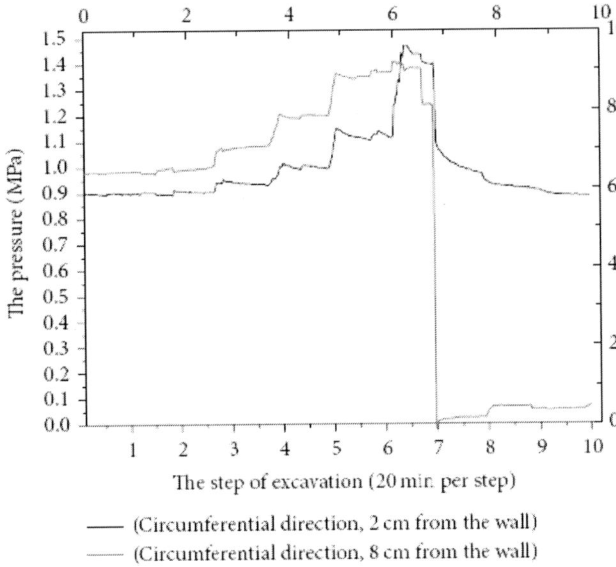

Figure 8. Hoop pressure change curve.

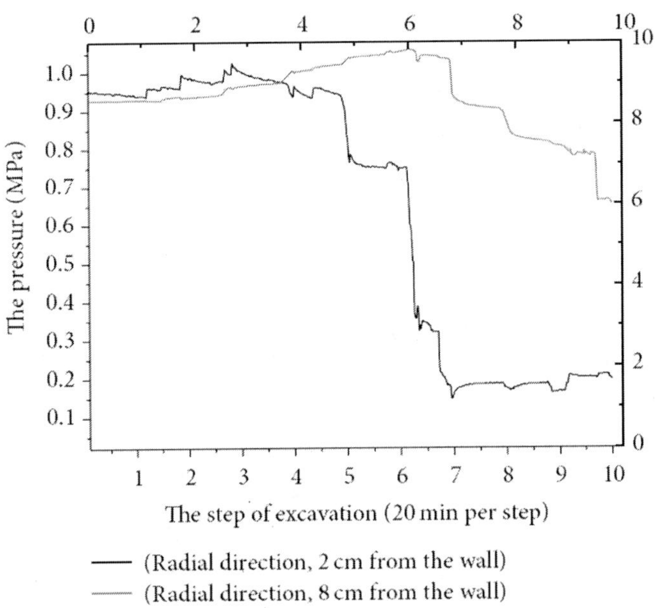

Figure 9. Radial pressure change curve.

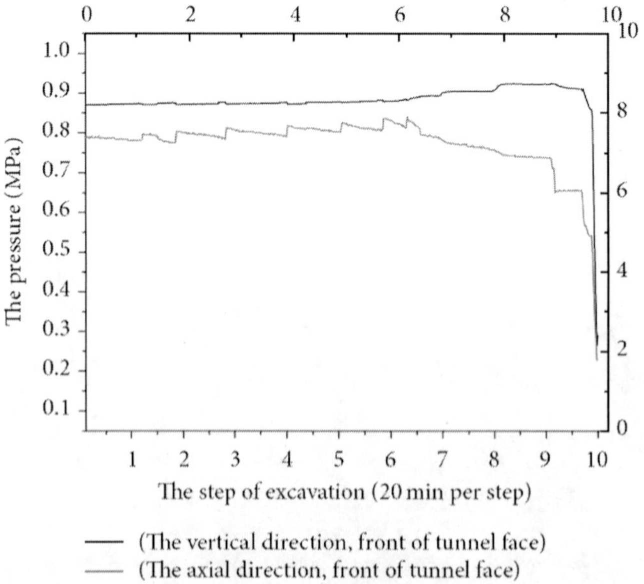

Figure 10. Internal pressure of measuring points (front of tunnel face).

The measurement results of circumferential pressure during the excavation are shown in Figure 8. It can be seen from the figure that the values of the hoop pressure adjust gradually with the process of excavation and unloading, showing a trend of gradual increase. When the excavation section was just at the position of the sensors, the pressure increased suddenly and rose to the maximum value. For example, for the measurement points 2 and 8 cm from the wall, the pressure increases from the original values of 0.9 and 0.98 MPa to 1.47 and 1.4 MPa, respectively. According to elastic theory, the stress growth amplitude is therefore 44% and 11%, respectively, which are less than the measurement values of 63.3% and 42.9%, respectively, in the same position of the test. This may be related to the high stress environment of deep rock mass or some mildly nonelastic behavior of the model material which results in a greater effective area. In addition, continuous excavation of the tunnel could also lead to a continual increase in pressure. When the hoop stress reaches the peak value after a certain period of time, the pressure falls sharply. The pressure 2 cm away from the wall decreases from 1.47 MPa to 0.89 MPa. The pressure 8 cm away from the wall suddenly decreases from 1.4 MPa to 0 MPa. This fact shows that the model at this point may be damaged or may have become loose due to possible local failure induced by extensional strain.

The pressure curve in the radial direction during the excavation is plotted in Figure 9. At the measurement points 2 and 8 cm from the wall, the pressure reduces from the original values of 0.96 and 0.92 MPa to 0.2 and 0.65 MPa, respectively. The amplitudes of the stress decrease for these two points are 79.2% and 29.3%, respectively. According to elastic theory, the amplitudes of the stress in these two points decrease 44% and 11%, respectively. The unloading of pressure in the radial direction induced by excavation is more apparent. The pressure at the point 2 cm away from the wall decreases sharply; this result shows that the model at this point may be damaged or become loose. So its pressure reduces to 0.2 MPa. In addition, the excavation process will lead to continuous loading along the circumferential direction and unloading along the radial direction in the surrounding rock mass. And the pressure variation is greater than the theoretical calculation, which is, therefore, prone to cause the failure of the surrounding rock and induce geological disasters.

The pressure curve in front of the tunnel face during the excavation is shown in Figure 10. There is little change in the pressure during the excavation. When the excavation section is close to the position of the sensors, the pressure in vertical direction decreases sharply from the original 0.87 MPa to 0.29 MPa, while the pressure in axial direction decreases gradually from the original 0.79 MPa to 0.23 MPa.

4.2. Strain Measurement

Two FBG strain sensors were embedded 2 cm from the wall along the radial and circumferential directions in the model. Figure 11 shows the strain curve of sensors during the process of test at each measuring point.

At the beginning stage of excavation, the two measurement points are in a relatively stable state. With the onset of excavation and unloading, the pressure of the surrounding rock gradually adjusts. The wavelength of the sensor in the hoop direction gradually decreases beginning with the third step of excavation. This is because of the increase in hoop stress due to excavation. The wavelength decreases the most at the sixth step of excavation. The maximum strain in hoop direction reaches 1268 $\mu\varepsilon$. On the other hand, the wavelength of the sensor in the radial direction suddenly increases during the fifth step of excavation. The

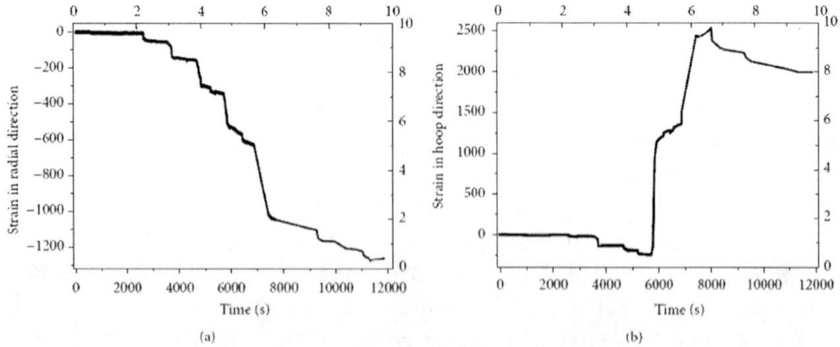

Figure 11. Internal strain change curve: (a) strain in radial direction; (b) strain in hoop direction.

corresponding strain reaches −1025 με. The maximum radial strain reaches −2528 με during the excavation. According to elastic theory, the radial and circumferential strains, which are 2 cm from the wall, are −1272 με and 1272 με, respectively. The measured variation rules of stress and strain are in good agreement with theoretical analysis, and the strain in the hoop direction is very close to the theoretical value.

4.3. Convergence of the Tunnel

Based on the small angle principle (Figure 12), the angle variation at the same point was measured by the Leica Total Station. The horizontal distance was measured from the marked points to the Total Station using tape. So the displacement can be calculated by [37]

$$\delta = \frac{\Delta\beta}{(\rho \cdot D)}, \tag{4}$$

where is angle variation between the two measurements, the constant coefficient denotes the relationship between a degree and a radian and is approximately equal to 206264.8, and is the distance between the measuring point and observation point.

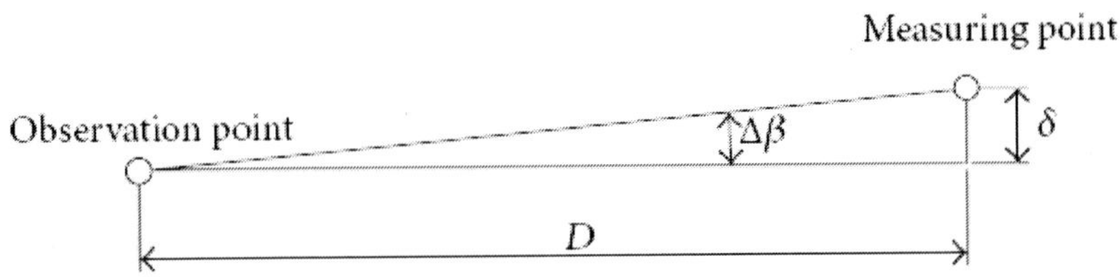

Figure 12. Displacement measuring principle of small angle method.

The displacement variations of the tunnel wall with time after excavation are shown in Figure 13. It is easy to see that the initial deformation of the tunnel grew steadily and then increased sharply, finally becoming stable. After 60 h, the displacements became stable, with convergences of 0.37, 0.53, and 0.55 mm at the roof, left-hand wall, and right-hand wall, respectively. The displacements corresponding to actual engineering are 27.75, 39.75, and 41.25 mm. The change rule of displacement in the tunnel agrees with the actual engineering excavation.

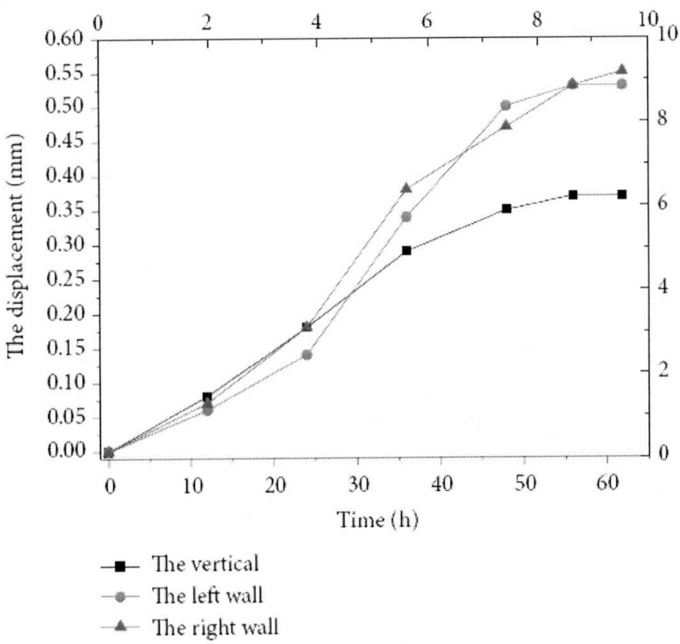

Figure 13. The convergence curve of tunnel wall after the excavation.

4.4. The Deformation and Failure of Surrounding Rock

In the construction of Jinping II hydropower project, the surrounding rock shows obvious brittle failure characteristics, such as plate crack phenomena and rockbursts. The occurrences of these kinds of failure phenomena are obvious during the experiment. Polyaxial or true triaxial tests had been performed by Addis et al. [38] to investigate the stability of model boreholes. And Zhou et al. [26] pointed out that there is a strong correlation between plate crack phenomena and rockbursts and that slab cracking is a precursor of rockbursts. In addition, the failure of the surrounding rock in the deep underground openings showed an obvious hysteresis effect. Obvious damage phenomena of the surrounding rock began to emerge 8 hours after excavation. The plate fracture failure of the surrounding rock occurred at the top and right side of the tunnel (Figure 14), which agrees with the practical engineering. Then two days after the excavation the ejection of fractured blocks into the hole occurred, which we interpreted as reproduction of a severe rock burst (Figure15).

Figure 14. The plate cracking phenomenon.

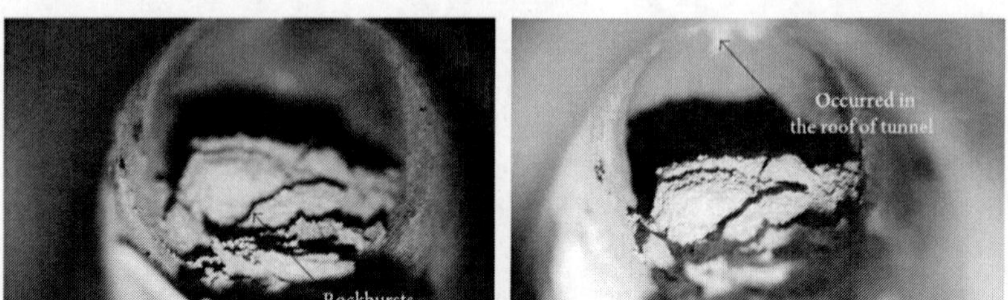

Figure 15. Rockbursts in model test.

From further analyses we can conclude that the stress level, under the deep rock mass environment, is usually very high. At the moment of creating a new opening, the increase in the hoop stress and the unloading of the radial stress result in a splitting tensile failure of the surrounding rock mass. Later, with further adjustment of the interior pressure and continual excavation and unloading, the surrounding rock gradually becomes damaged, and the peak of hoop stress transfers to the deep environment. After a certain period of time, the rocks surrounding the interior form a "support pressure zone." In the "range of the support pressure zone," the circumferential stress reaches the maximum value, while near the tunnel wall the surrounding rock loses a certain bearing capacity due to the gradual destruction. Thus, rockbursts suddenly occur 42 cm away from the model boundary. In the seventh step of excavation the pressure suddenly dropped to zero based on the measurement results, which indicated some damage was produced in the rock. However, rockbursts would not happen until a period of time after the excavation. With further adjustment of the surrounding rock pressure and potential energy concentration, rockbursts would occur suddenly. According to the ejected volume of equivalent material, this is a serious rockburst because the ejection volume is large. To sum up, a rockburst is a complex dynamic disaster phenomenon which occurs in the rock mass with high in situ stress induced by continual disturbances due to the excavation. The excavation brings a dramatic rise in the hoop stress and drop sharp in radial stress, which leads to the splitting failure of rock mass. Finally a strain model rockburst suddenly occurred associated with the release of strain energy stored in highly stressed rock mass.

5. CONCLUSIONS AND PROSPECTS

Through a preliminary test of a model for rockbursts simulated by low-strength brittle equivalent materials we conclude the following.

(1) Based on elastic adaptive loading boundary, Belleville springs are introduced to the loading apparatus. It is a new loading boundary of the test model which can minimize the boundary effect of geomechanical model test. By adjusting the spring system, the stiffness of test apparatus can be adjusted conveniently. It can quickly supply the model with a rebound energy caused by excavation and provide the model with a proper stiffness. Besides, stress can be adjusted automatically at the boundary.

(2) Owing to the excavation and unloading, there is a stress distribution and energy-dispersion process in the surrounding rock. At a distance of

2 and 8 cm from the tunnel wall, the radial pressure reduces from the original values 0.96 and 0.92 MPa to 0.2 and 0.65 MPa, respectively, and the hoop pressure increases from 0.9 and 0.98 MPa to 1.47 and 1.4 MPa, respectively. The maximum values of the radial and hoop strain reach to 2528 με and 1268 με during the process of the excavation, respectively. The convergences at the vertical, left-hand, and right-hand side of the wall in the tunnel after excavation were 0.37, 0.53, and 0.55 mm, respectively. The excavation process will lead to continuous loading along the hoop direction and unloading along the radial direction in the surrounding rock mass. The fact that pressure variation in the test is greater than the theoretical calculation more easily leads to the failure of surrounding rock.

(3) In the process of the test model, the brittle failure phenomenon of surrounding rock can be observed. The failure of the surrounding rock in deep underground openings demonstrated an obvious hysteresis effect. The mechanism of strain rockbursts can be interpreted as follows: with an initial state of high stress in rocks, the excavation brings a dramatic rise in the hoop stress and sharp drop in radial stress, which leads to the splitting failure of rock mass. Finally a strain model rockbursts occurred suddenly associated with the release of strain energy stored in highly stressed rock mass.

These results provide a basis for experiment and data to support the theoretical study of rockbursts. However, there is still some work to be done. The influence of stiffness of the test device on the destruction of the surrounding rock should be further investigated during excavation. To obtain more information of the dynamic mechanical parameters for rockbursts, the number of stress and strain measuring points should be increased.

COMPETING INTERESTS

The authors declare no competing interests.

ACKNOWLEDGMENTS

This study was financially supported by the National Basic Research Program of China (973 Program: 2013CB036005), the National Key Scientific Instrument and Equipment Development Project (no. 51527810), and the State Key Laboratory of Coal Resources and Safe Mining, CUMT (no. 14KF02).

REFERENCES

1. W. Blake, "Rock-burst mechanics," The Colorado School of Mines, vol. 67, no. 1, 1972.

2. M. Kurlenya, V. Oparin, and V. Vostrikov, "Effect of anomalously low friction in block media," Journal of Applied Mechanics and Technical Physics, vol. 40, no. 6, pp. 1116–1120, 1999.

3. Q. H. Qian, "New developments of rock engineering and technology in China," in Proceedings of the 12th ISRM International Congress on Rock Mechanics, Harmonising Rock Engineering and the Environment, pp. 57–61, Taylor and Francis Group, Beijing, China, October 2011.

4. E. I. Shemyakin, G. L. Fisenko, M. V. Kurlenya et al., "Zonal disintegration of rocks around underground workings, part 1: data of in situ observations," Soviet Mining, vol. 22, no. 3, pp. 157–168, 1986.

5. N. Cook, "The basic mechanics of rockbursts," Journal of the Southern African Institute of Mining and Metallurgy, vol. 63, pp. 71–81, 1963.

6. N. Cook, "A note on rockbursts considered as a problem of stability," Journal of the Southern African Institute of Mining and Metallurgy, vol. 65, no. 3, pp. 437–446, 1965.

7. M. D. G. Salamon, "Stability, instability and design of pillar workings," International Journal of Rock Mechanics and Mining Sciences & Geomechanics Abstracts, vol. 7, no. 6, pp. 613–631, 1970.

8. E. Hoek and E. T. Brown, Underground Excavations in Rock, Institution of Mining and Metallurgy, 1980.

9. A. Kidybiński, "Bursting liability indices of coal," International Journal of Rock Mechanics and Mining Sciences, vol. 18, no. 4, pp. 295–304, 1981.

10. J.-A. Wang and H. D. Park, "Comprehensive prediction of rockburst based on analysis of strain energy in rocks," Tunnelling and Underground Space Technology, vol. 16, no. 1, pp. 49–57, 2001.

11. A. V. Dyskin and L. N. Germanovich, "Model of rockburst caused by cracks growing near free surface,"Rockbursts and Seismicity in Mines, vol. 93, no. 1, pp. 69–75, 1993.

12. S. Nemat-Nasser and H. Horii, "Compression-induced nonplanar crack extension with application to splitting, exfoliation, and rockburst," Journal of Geophysical Research, vol. 87, no. 8, pp. 6805–6821, 1982.

13. H. Xie and W. G. Pariseau, "Fractal character and mechanism of rock bursts," International Journal of Rock Mechanics and Mining Sciences and, vol. 30, no. 4, pp. 343–350, 1993.

14. J. C. Gu, J. Fan, F. L. Kong, K. T. Wang, J. M. Xu, and T. Wang, "Mechanism of ejective rockburst and model testing technology," Chinese Journal of Rock Mechanics and Engineering, vol. 33, no. 6, pp. 1081–1089, 2014 (Chinese). 15.

15. S. P. Singh, "The influence of rock properties on the occurrence and control of rockbursts," Mining Science and Technology, vol. 5, no. 1, pp. 11–18, 1987.

16. M. Barquins and J.-P. Petit, "Kinetic instabilities during the propagation of a branch crack: effects of loading conditions and internal pressure," Journal of Structural Geology, vol. 14, no. 8-9, pp. 893–903, 1992.

17. P. S. Steif, "Crack extension under compressive loading," Engineering Fracture Mechanics, vol. 20, no. 3, pp. 463–473, 1984.

18. K. H. Höfer and K. Thoma, "Triaxial tests on salt rocks," International Journal of Rock Mechanics and Mining Sciences & Geomechanics Abstracts, vol. 5, no. 2, pp. 195–196, 1968.

19. H. B. Li, J. Zhao, and T. J. Li, "Triaxial compression tests on a granite at different strain rates and confining pressures," International Journal of Rock Mechanics and Mining Sciences, vol. 36, no. 8, pp. 1057–1063, 1999.

20. Y. Zhao, Z. Wan, Z. Feng, D. Yang, Y. Zhang, and F. Qu, "Triaxial compression system for rock testing under high temperature and high pressure," International Journal of Rock Mechanics and Mining Sciences, vol. 52, pp. 132–138, 2012.

21. M. C. He, J. L. Miao, and J. L. Feng, "Rock burst process of limestone and its acoustic emission characteristics under true-

triaxial unloading conditions," International Journal of Rock Mechanics and Mining Sciences, vol. 47, no. 2, pp. 286–298, 2010.

22. H. Manchao, X. Jia, M. Coli, E. Livi, and L. Sousa, "Experimental study of rockbursts in underground quarrying of Carrara marble," International Journal of Rock Mechanics and Mining Sciences, vol. 52, pp. 1–8, 2012.

23. M. He, D. Liu, W. L. Gong et al., "Development of a testing system for impact rock burst," Chinese Journal of Rock Mechanics and Engineering, vol. 33, no. 9, pp. 1729–1739, 2014 (Chinese).

24. Y.-J. Zuo, X.-B. Li, Z.-L. Zhou, C.-D. Ma, Y.-P. Zhang, and W.-H. Wang, "Damage and failure rule of rock undergoing uniaxial compressive load and dynamic load," Journal of Central South University of Technology (English Edition), vol. 12, no. 6, pp. 742–748, 2005.

25. J. Vacek and J. Chocholoušová, "Rock burst mechanics: insight from physical and mathematical modelling," Acta Polytechnica Scandinavica, vol. 48, no. 6, pp. 38–44, 2008.

26. H. Zhou, R. C. Xu, J. J. Lu, C. Q. Zhang, F. Z. Meng, and Z. Shen, "Study on mechanisms and physical simulation experiment of slab buckling rockburst in deep tunnel," Chinese Journal of Rock Mechanics and Engineering, vol. 34, supplement 2, pp. 3658–3666, 2015.

27. A.-H. Lu, X.-B. Mao, and H.-S. Liu, "Physical simulation of rock burst induced by stress waves," Journal of China University of Mining and Technology, vol. 18, no. 3, pp. 401–405, 2008.

28. T. B. Li, X. F. Wang, and L. B. Meng, "Physical simulation study of similar materials for rock burst,"Chinese Journal of Rock Mechanics and Engineering, vol. 30, no. 1, pp. 2610–2616, 2011 (Chinese).

29. L. W. Chen and S. W. Bai, "Research on plane strain model test of rockburst of circular cavern in hard brittle rockmass," Chinese Journal of Rock Mechanics and Engineering, vol. 26, no. 12, pp. 2504–2509, 2007.

30. M. C. Liao, Z. K. Guo, F. Liu, and W. X. Cheng, "Determination of the specimen size for model experiment on zonal fracturing of

deep rock mass," Journal of Disaster Prevention and Mitigation Engineering, vol. 26, no. 1, pp. 58–62, 2006 (Chinese).

31. M.-Y. Wang, Z.-P. Zhou, and Q.-H. Qian, "Tectonic, deformation and failure problems of deep rock mass," Chinese Journal of Rock Mechanics and Engineering, vol. 25, no. 3, pp. 448–455, 2006 (Chinese).

32. J. Haiming, F. Pengxian, W. Mingyang, and F. Xiang, "Elastic adaptive boundary for geomechanical model test," Electronic Journal of Geotechnical Engineering, vol. 19, pp. 2349–2357, 2014.

33. "The National Standards Compilation Group of People's Republic of China," Tech. Rep. GB/T1972–2005, Standards Press of China, Beijing, China, 2005 (Chinese).

34. L. Li, M. Y. Wang, P. X. Fan, Y. H. Cheng, Z. Z. Li, and H. M. Jiang, "Development of a deep underground engineering model experiment for loading and unloading test apparatus," Rock and Soil Mechanics, vol. 37, no. 1, pp. 297–304, 2016 (Chinese).

35. Q.-H. Qian, "Definition, mechanism, classification and quantitative forecast model for rockburst and pressure bump," Rock and Soil Mechanics, vol. 35, no. 1, pp. 1–6, 2014 (Chinese).

36. P. Fan, H. Xing, L. Ma et al., "Bulk density adjustment of resin-based equivalent material for geomechanical model test," Advances in Materials Science and Engineering, vol. 2015, Article ID 363869, 8 pages, 2015.

37. X. G. Zhu, "Application of small-angle method in deformation monitoring and research," Surveying and Mapping of Geology and Mineral Resources, vol. 27, no. 4, pp. 38–39, 2011.

38. M. A. Addis, N. Barton, S. C. Bandis, and J. P. Henry, "Laboratory studies on the stability of vertical and deviated boreholes," in Proceedings of the 65th Annual Technical Conference and Exhibition of the Society of Petroleum Engineers, New Orleans, La, USA, September 1990.

6

THE EFFECT OF SYNTHETIC HYDRATED CALCIUM ALUMINATE ADDITIVE ON THE HYDRATION PROPERTIES OF OPC

Jolanta Doneliene, Anatolijus Eisinas, Kestutis Baltakys, and Agne Bankauskaite

Department of Silicate Technology, Kaunas University of Technology, Radvilenu 19, LT-50270 Kaunas, Lithuania

ABSTRACT

The effect of synthetic CAH (130°C; 8 h; $CaO/(SiO_2 + Al_2O_3) = 0.55$; $Al_2O_3/(SiO_2 + Al_2O_3) = 0.1, 0.15$) with different crystallinity on the hydration kinetics of OPC at early stages of hydration was investigated. Also, the formation mechanism of compounds during OPC hydration was highlighted. It was determined that the synthetic CAH accelerated the initial reaction and shortened the induction period. Also, the second and third exothermic reactions begun earlier, and, during the latter reaction, the higher values of the heat flow were obtained in comparison with pure OPC samples. At later stages of hydration, synthetic CAH affect the OPC hydration as the usual pozzolanic additives; moreover, the larger values of cumulative heat were reached. It should be noted that the nature of synthetic CAH samples accelerated the dissolution of gypsum and stimulates the earliest C_3S hydration.

1. INTRODUCTION

Supplementary cementitious materials (SCMs), including fly ash, ground granulated blast furnace slag, silica fume, calcined clays, and natural pozzolans, are commonly blended with clinker to make Portland cement or are used as a partial replacement of this component in concrete [1–6]. The practice of using SCMs is increasing, with the world average percent clinker in cement having decreased from 85% in 2003 to 77% in 2010, and it is projected to further decrease to 71% in the future [7]. The use of SCMs is favorable to the industry, generally resulting in concrete with lower cost, lower environmental impact, higher long-term strength, and improved long-term durability [8–10].

It is known that SCMs can be of natural and synthetic origin or used as industrial by-products [11–20]. However, the application of natural minerals or industrial by-products in cement or concrete production is limited by the energy demanding processing of mentioned materials, which includes crushing, grinding, and size separation and, in some cases, it may also involve thermal activation. The latter process is applied for SCMs, which do not have or reach the required pozzolanic activity and also to alter their physical and chemical properties [11–13, 21].

For this reason, due to the high early strength, thermal and chemical stability, and excellent durability with respect to ordinary Portland cement (OPC) the synthetic SCMs have attracted a great scientific attention, when employed in building and ceramic industries [11, 22–29]. The one group of these compounds can be located in the $CaO–Al_2O_3–SiO_2–H_2O$ system. During hydrothermal synthesis the interreaction between silica, alumina, and calcium ions results in the formation of compounds such as calcium silicate hydrates (CSH), calcium aluminate hydrates (CAH), and calcium aluminium silicate hydrates (CASH). For this reason, many researchers keep their focus on tailoring the final structural and physical properties of mentioned materials mainly by changing the synthesis conditions (using different raw materials and changing ratio of primary mixture or the duration and temperature of isothermal curing) [30–35]. In Baltakys et al. work, when the duration of hydrothermal synthesis (175°C; 0–24 h) was extended, a lower amount of CASH was obtained [35]. Meanwhile, in Meller et al. work [31–34] the stability of obtained phases in $CaO–Al_2O_3–SiO_2–H_2O$ system was highlighted. It was found that the latter parameter and the formation mechanism of such compounds were significantly influenced by the synthesis temperature (from 200 to 350°C) and even small amounts of Al_2O_3additive.

In previous work [35] the influence of Al_2O_3 on the formation of calcium aluminium silicate hydrates under hydrothermal conditions was also examined. It was determined that the largest amount of calcium aluminium silicate hydrates was obtained after 8 h of hydrothermal treatment at 130°C, in the mixtures with a higher amount of Al_2O_3 ($Al_2O_3/(SiO_2 + Al_2O_3) = 0.15$). Meanwhile, the formation of mentioned compounds is inhibited, when the duration of isothermal curing is extended from 16 to 72 h.

Aluminium readily enters the calcium silicate hydrate phase (C–S–H) of Portland cement, and this substitution is expected to play a significant role in many aspects of the chemical behavior of cement paste, including the cation and anion exchange behavior, solubility, and the progress of the reactions that occur during delayed ettringite formation [36–42].

In previous work [35] it was proved that the synthetic CAH ($CaO/(SiO_2 + Al_2O_3) = 0.55$; $Al_2O_3/(SiO_2 + Al_2O_3) = 0.15$) as an additive affects the early hydration of OPC. This compound effectively shortens the induction period and accelerates the second and the third exothermic reactions.

For this reason, the main objective of this study was to determine the effect of synthetic CAH with different crystallinity on the hydration kinetics of OPC at early stages of hydration. Also, the compound formation mechanism is highlighted.

2. MATERIALS AND METHODS

2.1. Materials

In this paper the following reagents were used: $SiO_2 \cdot nH_2O$ ("Reaktiv," Russia) was grinded for 3 min in a vibrating cup "Pulverisette 9" mill (speed: 950 rpm) and sieved through a sieve with a mesh size of 80 μm, = 1047 m²/kg by CILAS LD 1090 granulometer, the loss of ignition, 7.0 wt.%. Calcium oxide from $Ca(OH)_2$("Reaktiv," Russia) was additionally burned at 500°C temperature for 2 h, grinded for 30 s in a vibrating cup "Pulverisette 9" mill (speed: 650 rpm), and sieved through a sieve with a mesh size of 80 μm; the quantity of free CaO is equal to 99 wt.%, = 961 m^2/kg. γ-Al_2O_3 was produced by burning aluminium hydroxide ("Sigma-Aldrich," Germany) at 475°C for 4 h, = 822 m^2/kg.

2.2. Synthetic CAH Samples Preparation

Dry primary mixtures with $CaO/(SiO_2 + Al_2O_3) = 0.55$ and $Al_2O_3/(SiO_2 + Al_2O_3) = 0.1$; 0.15 were mixed with water to reach the water/solid ratio of the suspension equal to 10.0. The hydrothermal synthesis has been carried out in unstirred suspensions in 25 mL volume PTFE cells, which were placed in "Parr instruments" (Germany) autoclave, under saturated steam pressure at 130°C temperature for 8 h (the temperature was reached within 2 h). After hydrothermal treatments, the autoclave was quenched to room temperature. The suspensions after synthesis were filtered and products were rinsed with ethanol to prevent carbonization of materials, dried at 50 ± 5°C temperature for 24 h, and sieved through a sieve with a width of 80 μm. These synthesis conditions were chosen according to previously published data [35].

2.3. Cementitious Admixtures with Synthetic CAH Samples

Samples of OPC were prepared in a laboratory grinding mill by grinding cement clinker (JSC "Akmenes cementas," Lithuania) with a 4.5% additive of gypsum ("Sigma-Aldrich," Germany) up to = 450 m^2/kg. The chemical analysis and phase composition of clinker are shown in Table 1. Synthetic CAH sample was added as a partial replacement of the OPC at levels of 10% by weight of the total cementitious material. Therefore the water and cement ratio (W/C) of all OPC samples was equal to 0.5.

Table 1: Chemical and mineralogical composition of clinker.

Oxides	SiO_2	Al_2O_3	Fe_2O_3	CaO	MgO	Na_2O	K_2O	SO_3^{2-}	Ignition of losses	Insoluble particles
Amount, %	19.72	5.41	4.21	62.76	3.41	0.16	1.08	2.08	0.93	0.24
Minerals	$3CaO{\cdot}SiO_2$		$2CaO{\cdot}SiO_2$			$3CaO{\cdot}Al_2O_3$			$4CaO{\cdot}Al_2O_3{\cdot}Fe_2O_3$	
Amount, %	63.19		8.89			7.21			12.81	

In the next stage of experiment, synthesized CAH samples were added as a partial replacement of the OPC at levels of 10% by weight of the total cementitious material. The additives of hydrated calcium aluminate were labeled as OPC-CAH1 (($CaO/(SiO_2 + Al_2O_3) = 0.55$, $Al_2O_3/(SiO_2 + Al_2O_3) = 0.1$)) and OPC-CAH2 ($CaO/(SiO_2 + Al_2O_3) = 0.55$, $Al_2O_3/(SiO_2 + Al_2O_3) = 0.15$) depending on the composition of primary mixtures.

2.4. Test Methods

The X-ray powder analysis (XRD) was performed on the D8 Advance diffractometer (Bruker AXS, Karlsruhe, Germany) operating at the tube voltage of 40 kV and tube current of 40 mA. The X-ray beam was filtered with Ni 0.02 mm filter to select the CuK_{α} wavelength. Diffraction patterns were recorded in a Bragg-Brentano geometry using a fast counting detector Bruker LynxEye based on silicon strip technology. The specimens were scanned over the range = 3–70° at a scanning speed of 6°C min^{-1} using a coupled two-theta/theta scan type.

Simultaneous thermal analysis (STA: differential scanning calorimetry, DSC, and thermogravimetry, TG) was also employed for measuring the thermal stability and phase transformation of samples at a heating rate of 15°C min^{-1}; the temperature ranged from 30°C up to 900°C under air atmosphere. The test was carried out on a Linseis instrument STA PT1000. The ceramic sample handlers and crucibles of Pt were used.

Differential scanning calorimetry (DSC) analysis was performed by a Netzsch DSC 214 Polyma instrument. This method was employed for measuring the thermal stability and phase transformation of samples at a heating rate of 10°C min^{-1}, and the temperature ranged from 30°C up to 600°C under air atmosphere. Ceramic sample handlers and Al crucibles were used.

An eight-channel TAM Air III isothermal calorimeter was used to investigate the heat evolution rate of the samples. Glass ampoules (20 mL) each containing 3 g dry cementitious material were placed in the calorimeter and the injection units for each ampoule filled with amounts of water equivalent to a W/(OPC and additive) ratio of 0.5. After a steady temperature of 25°C had been reached, the water was injected into the ampoules and mixed inside the calorimeter with the dry material for 20 s (frequency 2-3 s^{-1}). The heat evolution rate was then measured over a period of 72 h. Repetition of the measurements showed deviations in total heat below 3% for samples of similar type. Apart from the first minutes of water additive and mixing, the heat evolution rates were essentially identical. The rate of heat evolution was calculated on the basis of a unit weight of OPC.

In order to investigate the minerology and chemical composition of compounds formed during hydration, the heat evolution experiments were repeated at 25°C for different time periods (1.8, 3, 9.5, 13, and 24 h), which corresponded to the onset/peak/end duration of different early hydration periods. Hydration of samples was stopped by using acetone. Later on, the samples were crushed to powder, dried at the temperature of °C, and put through a sieve with an 80 μm mesh.

3. RESULTS AND DISCUSSION

3.1. Heat Evolution in Hydrated Cementitious Admixtures with Synthetic CAH Samples

The data of cumulative heat of hydration as well as the rate of heat evolution of the binary blended pastes are presented in Figure 1.

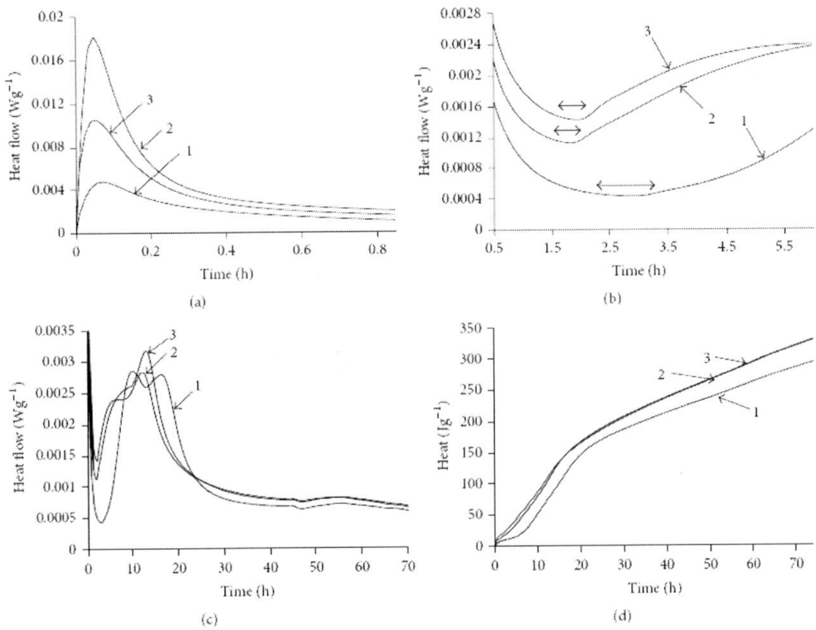

Figure 1. The heat evolution rate (a–c) and cumulative heat (d) of OPC (1), OPC-CAH1 (2), and OPC-CAH2 (3) samples during the early stage of hydration.

The heat of hydration curves for pure OPC and OPC with additives shows the typical five stages of the hydration reaction (the initial reaction, the induction period, the acceleratory period, the deceleratory period, and the period of slow continued reaction) as described in the literature [43–45].

It was determined that the additives of synthetic CAH samples in OPC samples accelerated the initial reaction (1-2 min) because an increase in the maximum heat evolution rate was observed from 0.005 Wg^{-1} to 0.011 Wg^{-1} (OPC-CAH2) and 0.018 Wg^{-1} (OPC-CAH1) (Figure 1(a)). And also, the induction period, which is assigned to the growing C-S-H and CH on the surface of the particles of primary compounds, was effectively shortened. In pure OPC samples, the mentioned process continued after 3 h, while in a case of OPC-CAH1 and OPC-CAH1 only after ~1.8 h (Figure 1(b)). However, the accelerating effect begun earlier: from 1.9 h in the samples with an additive and from 3.2 h in pure OPC samples (Figure 1(c)). Moreover, the lower values of the heat flow of the second exothermic reaction, typical to the dissolution of C_3S, were reached in OPC mixtures with additives: 0.0023 Wg^{-1} at 5.02 h for OPC-CAH1 and 0.0023 Wg^{-1} at 6.02 h for OPC-CAH2. Meanwhile, during the third exothermic reaction, which is characteristic to the dissolution of C_3A and the formation of ettringite, the highest value of the heat flow (0.0032 Wg^{-1}) at 13.1 h was obtained in OPC-CAH2 sample (Figure 1(c)).

At later stages of hydration, synthetic CAH samples affect the OPC hydration as the usual pozzolanic additive; moreover, the largest value of cumulative heat was reached in the OPC-CAH2 sample (Figure 1(d)).

3.2. Quantity of Compounds in the Cement Paste

It was determined that, during early OPC hydration, the stability and reactivity of OPC-CAH1 and OPC-CAH2 significantly depend on the primary mixture composition used for hydrothermal synthesis. OPC-CAH1 additive fully reacted already after 1.8 h of hydration; meanwhile, in a case of OPC-CAH2, only 33% (14.65 J/g) of calcium aluminium hydrate reacted (Figure 2(a): endothermic effect at ~271°C; Table 2).

Table 2: The main characteristics of thermal effects typical of CAH.

Composition	Hydration time (h)	Onset, °C	Peak, °C	Heat of process, J/g	Unreacted CAH, wt.%
OPC-CAH2	0	250.2	268.6	21.81	100
	1.8	253.8	271.0	14.65	67.17
	3	254.5	273.3	14.51	66.53
	5.5	256.7	276.3	13.67	62.68
	9.5	261.6	280.5	12.24	56.12
	13	265.2	282.5	11.94	54.75
	16.5	264.1	286.0	11.71	53.69
	24	267.7	286.8	10.39	47.64

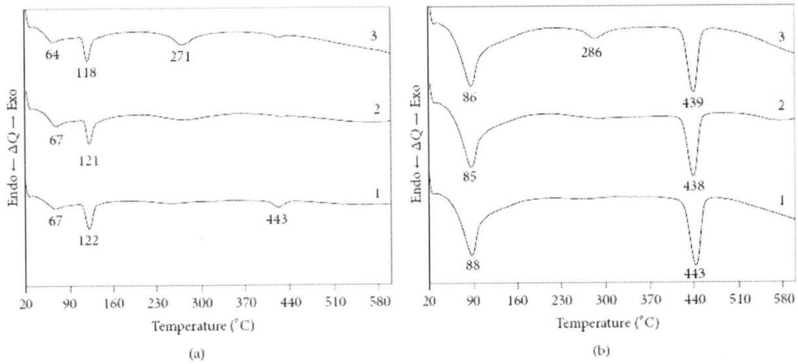

Figure 2. DSC curves OPC with 10 wt.% additive after 1.8 h (a) and 16.5 h (b): 1, OPC; 2, OPC-CAH1; 3, OPC-CAH2.

However, when the duration of OPC hydration was extended to 24 h, the amount of unreactive compounds was decreased (Figure 2; Table 2).

It should be noted that the nature of CAH accelerated the dissolution of gypsum. In the pure system, the latter compound fully reacted only after 16.5 hours of hydration, whereas in the samples with additives it fully reacted already after 13 h. It is clearly visible in DSC curve: the endothermic effect in a 105–135°C temperature range, which corresponded to the dehydration of gypsum, disappeared (Figure 2; Table 3). Moreover, in comparison with pure OPC samples, only a smaller amount of semicrystalline C-S-H (Figure 2: endothermic effect at 67°C; Table 3) as well as portlandite (Figure 2: endothermic effect at 440°C; Table 3) were obtained in the mixtures with additives by prolonging the duration of OPC hydration.

In this report, the amounts of C_3S that have reacted after heat evolution experiments at different time period under normal conditions were determined by the quantitative analysis (QXRD). The quantity of C_3S was calculated from the intensity change of the basic reflection (-spacing, 0.2604 nm). Each calculation was done five times, and it was determined that their data declined no more than ±2% from the mean (Figure 3). C_3S quantity in the mixtures of pure OPC and OPC with additives (before hydration) was equal to 100%.

Table 3. The main characteristics of thermal effects typical of CSH and gypsum by DSC method.

Composition	Hydration time (h)	CSH			Gypsum			
		Onset, °C	Peak, °C	Heat of process, J/g	Onset, °C	Peak, °C	Heat of process, J/g	Unreacted gypsum, wt.%
OPC	1.8	45.8	66.7	14.89	112.5	122.4	16.23	69.01
	3	48.0	64.4	17.22	108.1	118.3	16.69	70.96
	5.5	52.4	73.6	28.12	111.1	119.4	10.97	46.64
	9.5	54.7	76.9	40.88	113.3	119.8	6.86	29.15
	13	57.5	82.0	53.01	116.2	121.3	0.80	3.36
	16.5	61.0	87.8	71.84	—	—	—	—
	24	58.8	86.3	77.42	—	—	—	—
OPC-CAH1	1.8	49.0	66.8	18.61	113.6	121.0	13.63	74.93
	3	48.4	68.0	22.53	112.2	119.4	10.33	43.92
	5.5	54.5	75.7	37.60	115.1	121.0	6.02	25.57
	9.5	58.7	81.5	51.10	116.6	122.7	0.29	1.24
	13	62.7	87.6	63.13	—	—	—	—
	16.5	59.5	84.8	66.49	—	—	—	—
	24	58.5	84.5	71.74	—	—	—	—
OPC-CAH2	1.8	44.4	63.0	17.20	110.5	118.3	13.54	67.60
	3	49.4	69.3	21.10	111.7	119.6	11.15	55.67
	5.5	55.2	76.9	37.32	114.7	120.8	4.74	23.68
	9.5	64.7	85.1	54.86	116	121.8	0.51	2.54
	13	61.9	85.8	64.14	—	—	—	—
	16.5	61.4	85.1	58.51	—	—	—	—
	24	60.2	86.6	66.50	—	—	—	—

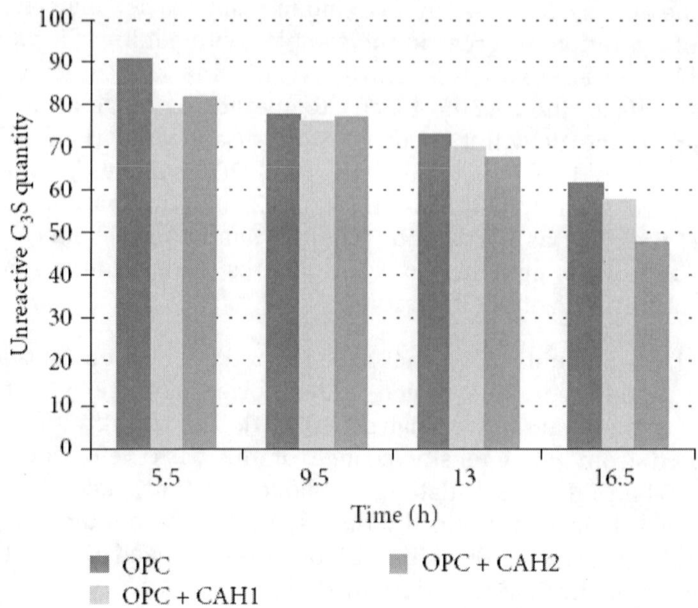

Figure 3. The quantity of unreactive C_3S in pure OPC, OPC-CAH1, and OPC-CAH2 samples after different duration of hydration.

It was determined that after 5.5 h of hydration, only 9% of C_3S reacts in pure OPC samples and the further reduction of its quantity depends on duration of hydration (Figure 3). Meanwhile, within the same duration of hydration, in the samples with CAH1 and CAH2 additives, 21% and 18% of this compound reacted, respectively (Figure 3). It was noticed that mentioned additives stimulate the earliest C_3S hydration in all samples (5.5–16.5 h) (Figure 3).

Presumably, the synthetic CAH sample also induced the formation mechanism of ettringite.

In order to prove this fact, the intensity change of the basic reflection of ettringite (-spacing, 0.972 nm) in cement samples was evaluated by X-ray diffraction analysis (XRD) according to the area of mentioned diffraction maximum per shift at different hydration time.

The obtained results showed that after 3 hours of hydration, the area of the main diffraction peak typical to ettringite increased in two times in comparison with the pure OPC samples (Figure 4). The same tendency was also observed, when the duration of hydration was extended to 9.5 hours (Table 4).

Table 4: The values of the main diffraction peak area of ettringite in OPC, OPC-CAH1, and OPC-CAH2 samples.

Composition	Hydration time (h)							
	0	1.8	3	5.5	9.5	13	16.5	24
OPC	—	0.14	0.15	0.26	0.40	0.60	1.00	1.45
OPC-CAH1	—	0.15	0.37	0.42	0.89	0.90	1.18	1.39
OPC-CAH2	—	0.21	0.22	0.54	0.51	1.03	1.18	1.30

In addition, due to the accelerated dissolution of gypsum, a larger amount of formed ettringite was observed in OPC-CAH1 and OPC-CAH2 samples. Meanwhile, in an excess of gypsum in the pure OPC samples, the area of the main diffraction maximum of ettringite was slightly higher after 24 h of hydration (Table 4).

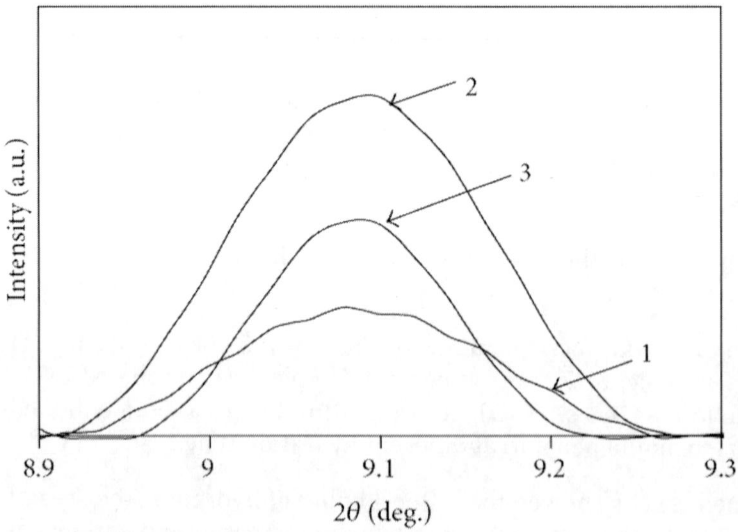

Figure 4. The change of the main diffraction peak of ettringite in samples after 3 h of hydration: 1, OPC; 2, OPC-CAH1; 3, OPC-CAH2.

4. CONCLUSIONS

It was examined that the additives accelerated the initial reaction and shortened the induction period. Also, the second and third exothermic reactions begun earlier, and, during the latter reaction, the higher values of the heat flow were obtained in comparison with pure OPC samples. At later stages of hydration, synthetic CAH affect the OPC hydration as the usual pozzolanic additives; moreover, the larger values of cumulative heat were reached.

It was noticed that additives stimulate the earliest C_3S hydration (5.5–16.5 h): after 5.5 h of hydration, only 9% of C_3S reacts in pure OPC samples and the further reduction of its quantity depends on duration of hydration. Meanwhile, in a case of CAH1 and CAH2 additives, 21% and 18% of this compound reacted.

It should be noted that the nature of synthetic CAH samples accelerated the dissolution of gypsum. In the pure system, the latter compound fully reacted only after 16.5 hours of hydration, whereas in the samples with additives it reacted already after 13 h. For this reason, a larger amount of formed ettringite was observed in OPC-CAH1 and OPC-CAH2 samples till 16.5 h of hydration.

COMPETING INTERESTS

The authors declare that there is no conflict of interests regarding the publication of this paper.

ACKNOWLEDGMENTS

This research was funded by a grant (no. MIP-025/2014) from the Research Council of Lithuania.

REFERENCES

1. C. H. Un, J. G. Sanjayan, R. San Nicolas, and J. S. J. van Deventer, "Predictions of long-term deflection of geopolymer concrete beams," Construction and Building Materials, vol. 94, pp. 10–19, 2015.

2. E. Gartner and H. Hirao, "A review of alternative approaches to the reduction of CO_2 emissions associated with the manufacture of the binder phase in concrete," Cement and Concrete Research, vol. 78, pp. 126–142, 2015.

3. K.-H. Yang, Y.-B. Jung, M.-S. Cho, and S.-H. Tae, "Effect of supplementary cementitious materials on reduction of CO_2 emissions from concrete," Journal of Cleaner Production, vol. 103, pp. 774–783, 2015.

4. E. Aprianti, P. Shafigh, S. Bahri, and J. N. Farahani, "Supplementary cementitious materials origin from agricultural wastes—a review," Construction and Building Materials, vol. 74, pp. 176–187, 2015.

5. S. H. Kosmatka, B. Kerkhoff, and W. C. Panarese, "Fly ash, slag, silica fume, and natural pozzolans," inDesign and Control of Concrete Mixtures, EB001, chapter 3, Portland Cement Association, 14th edition, 2008.

6. M. H. Cornejo, J. Elsen, C. Paredes, and H. Baykara, "Thermomechanical treatment of two Ecuadorian zeolite-rich tuffs and their potential usage as supplementary cementitious materials," Journal of Thermal Analysis and Calorimetry, vol. 115, no. 1, pp. 309–321, 2014.

7. M. C. G. Juenger and R. Siddique, "Recent advances in understanding the role of supplementary cementitious materials in concrete," Cement and Concrete Research, vol. 78, pp. 71–80, 2015.

8. A. Korpa, T. Kowald, and R. Trettin, "Hydration behaviour, structure and morphology of hydration phases in advanced cement-based systems containing micro and nanoscale pozzolanic additives," Cement and Concrete Research, vol. 38, no. 7, pp. 955–962, 2008.

9. N. M. Khalil, E. M. Hassan, M. M. E. Shakdofa, and M. Farahat, "Beneficiation of the huge waste quantities of barley and rice husks as well as coal fly ashes as additives for Portland cement," Journal of Industrial and Engineering Chemistry, vol. 20, no. 5, pp. 2998–3008, 2014.

10. M. Hamidi, L. Kacimi, M. Cyr, and P. Clastres, "Evaluation and improvement of pozzolanic activity of andesite for its use in eco-efficient cement," Construction and Building Materials, vol. 47, pp. 1268–1277, 2013.

11. M. S. Amin, S. A. Abo-El-Enein, A. A. Rahman, and K. A. Alfalous, "Artificial pozzolanic cement pastes containing burnt clay with and without silica fume: physicochemical, microstructural and thermal characteristics," Journal of Thermal Analysis and Calorimetry, vol. 107, no. 3, pp. 1105–1115, 2012.

12. B. Lothenbach, K. Scrivener, and R. D. Hooton, "Supplementary cementitious materials," Cement and Concrete Research, vol. 41, no. 12, pp. 1244–1256, 2011.

13. C. A. Utton, M. Hayes, J. Hill, N. B. Milestone, and J. H. Sharp, "Effect of temperatures up to 90°C on the early hydration of Portland-Blastfurnace slag cements," Journal of the American Ceramic Society, vol. 91, pp. 948–954, 2008.

14. L. Krishnan, S. Karthikeyan, S. Nathiya, and K. Suganya, "Geopolymer concrete an eco-friendly construction material," International Journal of Research in Engineering and Technology, vol. 3, pp. 164–167, 2014.

15. M. E. Natali, S. Manzi, C. Chiavari, and M. C. Bignozzi, "Steel reinforced geopolymer mortar: corrosion behaviour in chloride-rich environment," in Proceedings of the 34th Cement and

Concrete Science Conference, S. A. Bernal and J. L. Provis, Eds., pp. 41–44, University of Sheffield, September 2014.

16. A. Rafeet, R. Vinai, W. Sha, and M. Soutsos, "Alkali activated fuel ash and slag mixes: optimization study from mortars to concrete building blocks," in Proceedings of the 34th Cement and Concrete Science Conference, S. A. Bernal and J. L. Provis, Eds., pp. 349–354, University of Sheffield, September 2014.

17. C. Villa, E. T. Pecina, R. Torres, and L. Gómez, "Geopolymer synthesis using alkaline activation of natural zeolite," Construction and Building Materials, vol. 24, no. 11, pp. 2084–2090, 2010.

18. M. Babaee, A. Castel, and A. Akbarnezhad, "Active steel corrosion in blended slag and fly ash geopolymer concrete," in Proceedings of 34th Cement and Concrete Science Conference, S. A. Bernal and J. L. Provis, Eds., pp. 55–58, University of Sheffield, Sheffield, UK, 2014.

19. T. Bakharev, "Geopolymeric materials prepared using Class F fly ash and elevated temperature curing,"Cement and Concrete Research, vol. 35, no. 6, pp. 1224–1232, 2005.

20. T. Bakharev, "Resistance of geopolymer materials to acid attack," Cement and Concrete Research, vol. 35, no. 4, pp. 658–670, 2005.

21. P. K. Mehta and P. J. M. Monteiro, Concrete: Microstructure, Properties, and Materials, The McGraw-Hill Companies, 3rd edition, 2005.

22. G. Roviello, L. Ricciotti, C. Ferone, F. Colangelo, and R. Cioffi, "Hybrid organic-inorganic materials: novel perspectives for the application of geopolymer based materials," in Proceedings of the 34th Cement and Concrete Science Conference, S. A. Bernal and J. L. Provis, Eds., pp. 199–202, University of Sheffield, 2014.

23. K. Baltakys, A. Eisinas, T. Dizhbite, L. Jasina, R. Siauciunas, and S. Kitrys, "The influence of hydrothermal synthesis conditions on gyrolite texture and specific surface area," Materials and Structures, vol. 44, no. 9, pp. 1687–1701, 2011.

24. K. Baltakys, R. Siauciunas, R. Gendvilas, A. Eisinas, T. Dambrauskas, and S. Kitrys, "Physically and chemically bound

H_2O in the α-C_2S hydrate structure," Journal of Thermal Analysis and Calorimetry, vol. 118, no. 2, pp. 807–816, 2014.

25. H. C. Yong, H. Kamarudin, M. M. Al Bakri Abdullah, and L. Y. Ming, "Effect of alkali concentration on mechanical properties of kaolin geopolymers," Romanian Journal of Materials, vol. 42, no. 2, pp. 179–186, 2012.

26. M. A. Longhi, F. Gaedke, E. D. Rodríguez, A. C. Passuello, and A. P. Kirchheim, "Geopolymers based on calcined kaolin sludge/bottom ash blends and an alternative sodium silicate activator," in Proceedings of 34th Cement and Concrete Science Conference, S. A. Bernal and J. L. Provis, Eds., pp. 359–362, University of Sheffield, 2014.

27. B. Walkley and J. S. J. van Deventer, "Stoichiometrically controlled C-A-S-H / N-A-S-H gel blends via alkali-activation of synthetic precursors," in Proceedings of the 34th Cement and Concrete Science Conference, S. A. Bernal and J. L. Provis, Eds., pp. 27–32, University of Sheffield, Sheffield, UK, 2014.

28. K. Baltakys, T. Dambrauskas, R. Siauciunas, and A. Eisinas, "The formation of α-C_2S hydrate in the mixtures with $CaO/SiO_2 = 1.75$ by hydrothermal treatment at 200°C," Revista Romana de Materiale—Romanian Journal of Materials, vol. 44, pp. 109–155, 2014.

29. G. Mucsi, Z. Molnár, Á. Rácz, R. Szabó, and B. Csőke, "Geopolymer from mechanically activated low and high calcium fly ash," in Proceedings of the 34th Cement and Concrete Science Conference, S. A. Bernal and J. L. Provis, Eds., pp. 1391–1396, University of Sheffield, Sheffield, UK, 2014.

30. R. Siauciunas and A. Baltusnikas, "Influence of SiO_2 modification on hydrogarnets formation during hydrothermal synthesis," Cement and Concrete Research, vol. 33, no. 11, pp. 1789–1793, 2003.

31. N. Meller, C. Hall, K. Kyritsis, and G. Giriat, "Synthesis of cement based $CaO–Al_2O_3–SiO_2–H_2O$ (CASH) hydroceramics at 200 and 250 °C: ex-situ and in-situ diffraction," Cement and Concrete Research, vol. 37, no. 6, pp. 823–833, 2007.

32. N. Meller, K. Kyritsis, and C. Hall, "The mineralogy of the CaO-Al_2O_3-SiO_2-H_2O (CASH) hydroceramic system from 200 to 350

°C," Cement and Concrete Research, vol. 39, no. 1, pp. 45–53, 2009.

33. N. Meller, C. Hall, and J. S. Phipps, "A new phase diagram for the $CaO–Al_2O_3–SiO_2–H_2O$ hydroceramic system at 200 °C," Materials Research Bulletin, vol. 40, no. 5, pp. 715–723, 2005.

34. N. Meller, K. Kyritsis, and C. Hall, "The hydrothermal decomposition of calcium monosulfoaluminate 14-hydrate to katoite hydrogarnet and β-anhydrite: an in-situ synchrotron X-ray diffraction study,"Journal of Solid State Chemistry, vol. 182, no. 10, pp. 2743–2747, 2009.

35. K. Baltakys, A. Eisinas, J. Doneliene, D. Monstvilaite, A. Bankauskaite, and A. Urbutis, "The hydrothermal synthesis of calcium aluminium silicate hydrates and its application on Portland cement hydration," Romanian Journal of Materials, vol. 45, no. 3, pp. 218–225, 2015.

36. C. A. Ríos, C. D. Williams, and M. A. Fullen, "Hydrothermal synthesis of hydrogarnet and tobermorite at 175 °C from kaolinite and metakaolinite in the $CaO-Al_2O_3-SiO_2-H_2O$ system: a comparative study,"Applied Clay Science, vol. 43, no. 2, pp. 228–237, 2009.

37. I. G. Richardson, "The nature of C–S–H in hardened cements," Cement and Concrete Research, vol. 29, no. 8, pp. 1131–1147, 1999.

38. I. G. Richardson and J. G. Cabrera, "Nature of C–S–H in model slag-cements," Cement and Concrete Composites, vol. 22, no. 4, pp. 259–266, 2000.

39. A. R. Brough, A. Katz, G.-K. Sun, L. J. Struble, R. J. Kirkpatrick, and J. F. Young, "Adiabatically cured alkali-activated cement-based wasteforms containing high levels of fly ash, formation of zeolites and Al-substituted C–S–H," Cement and Concrete Research, vol. 31, pp. 1437–1447, 2001.

40. J. Schneider, M. A. Cincotto, and H. Panepucci, "29Si and 27Al high-resolution NMR characterization of calcium silicate hydrate phases in activated blast-furnace slag pastes," Cement and Concrete Research, vol. 31, no. 7, pp. 993–1001, 2001.

41. O. P. Shrivastava and R. Shrivastava, "Sr^{2+} uptake and leachability study on cured aluminum-substituted tobermorite-

OPC admixtures," Cement and Concrete Research, vol. 31, no. 9, pp. 1251–1255, 2001.

42. S.-Y. Hong and F. P. Glasser, "Alkali sorption by C–S–H and C–A–S–H gels: part II. Role of alumina,"Cement and Concrete Research, vol. 32, no. 7, pp. 1101–1111, 2002.

43. D. Jansen, F. Goetz-Neunhoeffer, B. Lothenbach, and J. Neubauer, "The early hydration of Ordinary Portland Cement (OPC): an approach comparing measured heat flow with calculated heat flow from QXRD," Cement and Concrete Research, vol. 42, no. 1, pp. 134–138, 2012.

44. P. Juilland, E. Gallucci, R. Flatt, and K. L. Scrivener, "Dissolution theory applied to the induction period in alite hydration," Cement and Concrete Research, vol. 40, no. 6, pp. 831–844, 2010.

45. A. Kumar, S. Bishnoi, and K. L. Scrivener, "Modelling early age hydration kinetics of alite," Cement and Concrete Research, vol. 42, no. 7, pp. 903–918, 2012.

7

SYNTHESIS AND CHARACTERIZATION OF ADVANCED RED MUD AND MWCNTS BASED EMI SHIELDING MATERIAL VIA CERAMIC PROCESSING

Sarika Verma*, Sudhir Sitaram Amritphale, Satyabrata Das

Council of Scientific and Industrial Research-Advanced Materials and Processes Research Institute, Bhopal, India

ABSTRACT

For the first time in the world advanced multi layered Red Mud and MWCNTs (ARMC) based EMI shielding material has been developed at CSIR-AMPRI, Bhopal. Red mud provides oxides of titanium and iron as precursor and the MWCNTs provides electrical conductivity characteristics necessary for making desired EMI shielding materials. The novel process involves unique designing of chemical compositions and mineralogical phases of red mud, MWCNTs together with appropriate additive and solvent which results in the simultaneous and synergistic chemical reactions among various constituents thereby forming tailored precursor powder. Further, the ceramic processing of tailored precursor powder in appropriate environment enables formation of advanced ARMC shielding material having a variety of ceramic phases with multi elemental compositions and multi layered crystal structures. The synthesized material was characterized by various techniques namely XRD, PL, FESEM, EDXA. The reflection loss (R. L.) of the sample was calculated based on the measured complex permittivity and permeability. The advanced ARMC material with thickness t = 1.5 mm showed a minimum R. L. of −35.5 Db at 14.0 GHz with a response band width of 1.8 GHz. Thus, the developed advanced ARMC material acts as a good EMI wave absorber.

Keywords: Red Mud, MWCNTs, EMI Shielding, Ceramic Processing, Multilayered Structure

1. INTRODUCTION

The present era is known as the era of radiation based technologies in all walk of the human race. Electromagnetic shielding is reducing the electromagnetic field by blocking it with materials having either conductive, magnetic or both the characteristics. Shielding of Electromagnetic Interference (EMI) is in demand 1) due to the interference of wireless devices with digital devices, 2) due to the increasing sensitivity, and 3) due to increased importance of electronic devices [1] - [4] . Shielding is particularly required for power transformers containing vaults and other electronics that are related to electric power and telecommunications. The desired characteristics for obtaining advanced EMI shielding materials should preferably be a) light in weight, b) efficiently absorbing EM waves, c) possessing tunable absorption frequency, d) multifunctional, e) resistance to corrosion, f) flexible and g) having processing advantages [5] - [9] .

Conventional EMI shielding materials are made using synthetic grade compounds of barium, titanium, iron and MWCNTs. The chemical reaction among these compounds leads to the formation of barium titanate, iron oxides in MWCNTs composite matrix. As reported by Tan et al. (2010) and Iyer et al. (1993), $BaTiO_3$ and TiO_2 posses high permittivity and dielectric constant, thereby improving the EMI shielding properties of the materials [10] [11] . Also as reported by Azadmanjiri et al. (2010) and Zhiguang Ding et al. (2015), the magnetic phases due to Fe_2O show a marked increase in the absorption of EMI waves [12] [13] .

In the present novel process red mud has been explored as a source of oxides of titanium and iron as precursor, MWCNTs provides desired electrical conductivity characteristics and barium hydroxide, being a source of barium for making advanced EMI shielding material. Red mud [14] is an aluminum industry waste which inherently contains varieties of elements namely iron, titanium, aluminum, silicon, calcium, magnesium and sodium etc and thus makes it a most suitable multi-component resource material for developing value added materials. Further these multi-elements in the red mud are present in the form of oxide, oxy-hydroxide and hydroxides, having low as well as high atomic numbers elements namely sodium, iron compounds respectively.

The unique structure and excellent properties of Carbon Nano Tubes (CNTs) (both single-walled and multi- walled (SWCNTs and MWCNTs)) like high aspect ratio, high conductivity, small diameter and mechanical strength have prompted its potential engineering applications, in the area of electronics, automotive and aerospace sectors. Further CNTs are use in electrostatic dissipation, electromagnetic interference (EMI) shielding, multilayer printed circuits and conductive coatings, etc. Several studies of using MWCNTs, as excellent option, for creating conductive composites for high-performance EMI shielding materials, have been reported [15] - [19] .

So far no serious efforts have been made to explore the iron and titanium precursors of Red mud and MWCNTs together, for the development of novel advanced EMI shielding material. In order to obtain particle size of developed advanced ARMC shielding material ranging from micron to submicron up to nano size, cetyltrimethylammonium bromide has been used for reducing the particles size down to the nanometer range as well as stabilizing the formed particles [20] [21] .

2. EXPERIMENTAL

2.1. Material and Chemicals Required

2.1.1. Red Mud

The Red Mud was collected from HINDALCO, Renukoot (U.P) for the study was analyzed in the laboratory for physico-chemical characteristics and the results are given in Table 1.

2.1.2. Chemicals

Multiwalled CNTs, Barium hydroxide, Sodium hydroxide, Cetyltrimethylammonium bromide (CTAB), ammonium nitrate were procured from Rankem and Ethylene glycol was procured from Merck. All the chemicals were used as such without further purification.

2.1.3. Procedure

The experimental work for the obtaining advanced ARMC shielding material has been divided into two parts in the following way.

1) Synthesis of tailored precursor powder

Synthesis of tailored precursor powder was done by together refluxing of 10 g of Red mud, 10 g barium hydroxide, 10 g MWCNTs, 15 g of cetyltrimethylammonium bromide and 20 g of sodium hydroxide in 400 ml of ethylene glycol for the duration of 4 h. The product was then precipitated with 30 ml of aqueous ammonium nitrate solution (25 wt %) and washed with the same solution to ensure complete precipitation. The precipitate so obtained was dried in air oven at 110°C for 2 h and grayish black tailored precursor powder was obtained.

Table 1. Chemical analysis of red mud HINDALCO, Renukoot (U.P).

Sr. No.	Constituents	%
1	LOI	8 - 10
2	Al_2O_3	17 - 19
3	Fe_2O_3	34 - 40
4	SiO_2	7 - 8
5	TiO_2	15 - 16
6	Na_2O	5 - 6
7	CaO	1.8 - 5.0
8	P_2O_5	0.31 - 0.47
9	V_2O_5	0.035

2) Ceramic processing of tailored precursor powder to obtain advanced ARMC shielding material

The resulting tailored precursor powder so obtained was converted in the form of rectangular coupon using warm press and was kept in furnace for sintering at 800°C for two hours in argon atmosphere i.e.

ceramic processing of tailored precursor to obtain EMI shielding material.

The various steps involved in the developed novel process for synthesis of advanced ARMC shielding materials shown in Figure 1.

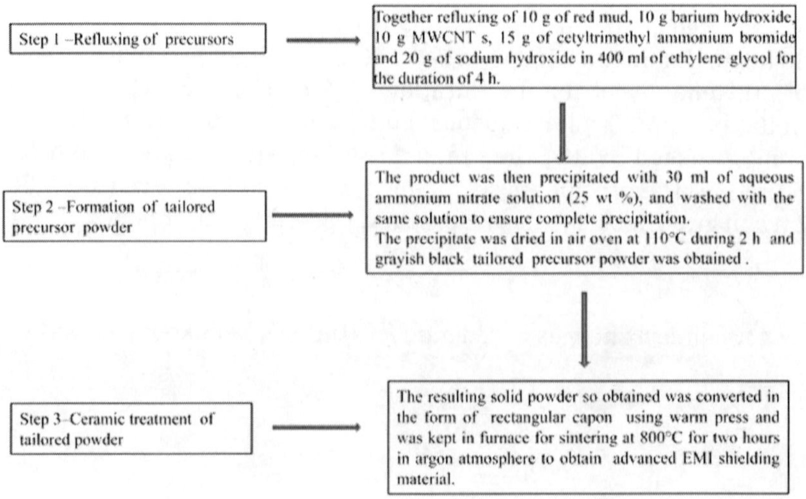

Figure 1. Flow sheet of advanced ARMC shielding material.

2.2. Characterization

X-ray diffraction pattern was obtained on D8 advance X-ray diffractometer using Cu K_α radiation to identify the various different phases. The X-ray diffraction intensity was recorded as a function of Bragg's 2θ in the angular range of 5° - 70°. The Photo luminescence measurement was recorded in Instrument Model: F-7000 FL Spectrophotometer with Serial Number: 2249-002 over the wavelength range from 300 nm to 700 nm. Field Emission Scanning Electron Microscope (FESEM), model NOVA NANOSEM-430 of COMFEI and Energy-disper- sive X-ray spectroscopy (EDXA), Model X-MAX of Oxford was used for determination of morphology and microstructure. Sample was sonicated for a period of 15 minutes in acetone before mounting for FESEM and EDXA.

3. RESULTS AND DISCUSSION

3.1. X-Ray Diffraction Studies

Identification of the various phases present in the synthesized advanced EMI shielding material formed during the reaction process was carried out by comparing the experimental inter planar spacing (d values) with those of the respective likely substances listed in the JCPDS standard X-ray diffraction (XRD) data files [22] . The X-ray diffraction spectrum (XRD) of the synthesized EMI material is shown in Figure 2 and the respective phases identified are given in Table 2 respectively.

In general the XRD patterns are composed of sharp peaks, indicating that the synthesized materials have good crystalline structure. The results of X-ray powder diffraction analysis of advanced material exhibited the presence of a diverse mineralogical phases and compounds namely barium iron titanium silicate (Bafertisite), barium aluminum silicate, iron titanium oxide, barium titanate, barium iron titanium oxide, barium aluminum oxide, magnetite, iron sulphide (troilite).

Crystallite size was obtained by using the Debye's Scherrer equation [23] :

$$D = \frac{K}{\beta \cos(\theta)}$$

where K is the shape factor, λ is the wavelength, β is the line broadening at half the maximum intensity (FWHM) in radians and θ is the Bragg angle. The crystallite size was of the synthesized material was found to be in the range from 10 - 18 nm.

3.2. Photoluminescence

The Photoluminescence spectra of the developed advanced ARMC shielding material is shown inFigure 3 and shows multiple peaks from 300 to 700 nm which confirms the presence of multi elemental and multi phased characteristics in the material. As reported by Ming-sheng Zhang et al. Strong photoluminescence spectrum centered at 696 nm was observed confirming the presence of barium titanate in the sample [24] .

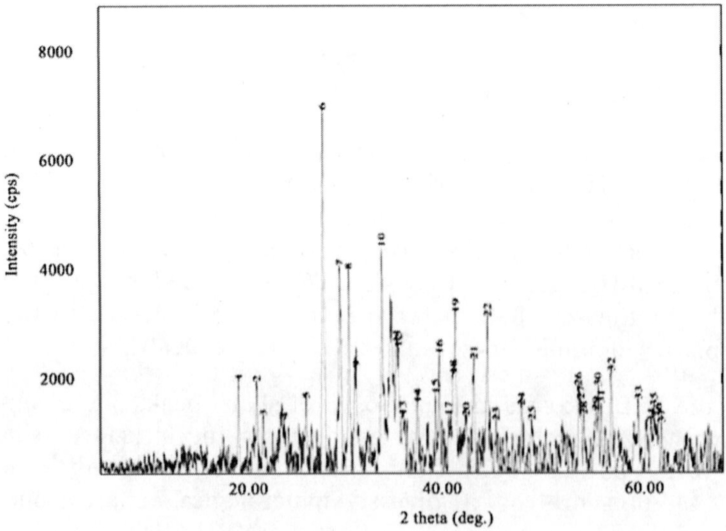

Figure 2. XRD pattern of advanced ARMC shielding material.

Table 2. Phases present in advanced ARMC shielding material.

No.	Phase identified in material	JCPDS data (file number)	Crystal structure
1	Barium iron titanium silicate (Bafertisite)	14 - 541	Orthorhombic
2	Barium aluminum silicate	28 - 125	Hexagonal
3	Iron titanium oxide (pseudorutile)	13 - 270 and 19 - 635	Tetragonal and hexagonal
4	Barium titanate	8 - 368	Monoclinic
5	Barium iron titanium oxide	26 - 1032	Hexagonal
6	Barium aluminum oxide	28 - 121	Tetragonal to cubic
7	Magnetite	19 - 629	Cubic
8	Iron sulphide (troilite)	11 - 151	Hexagonal phases

3.3. FESEM and EDXA Analysis

The particle size of synthesized advanced AFMC shielding material was examined using FESEM and EDXA respectively. Several particles were investigated to determine the particle size. FESEM image and EDXA spectra are shown in Figure 4 and Figure 5 respectively.

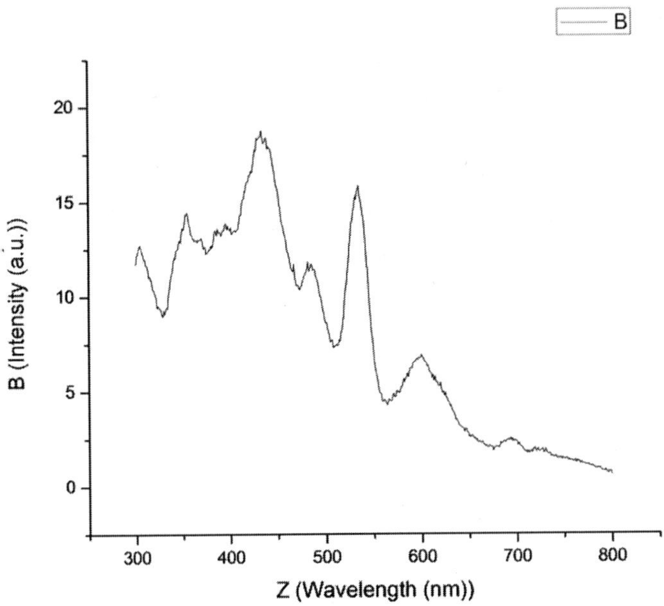

Figure 3. PL of advanced ARMC shielding material.

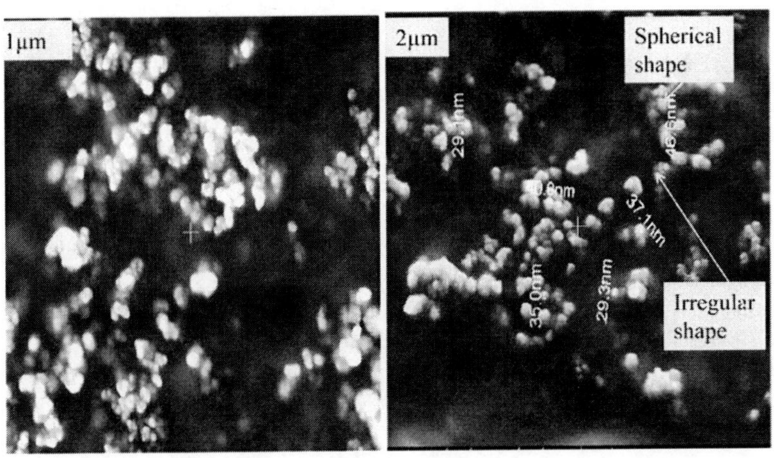

Figure 4. FESEM image of of advanced ARMC shielding material.

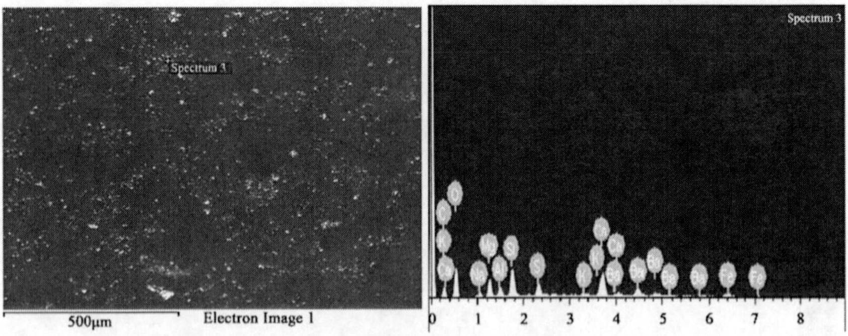

(a) (b)

Element	Weight%	Atomic%
C K	16.00	34.24
O K	25.61	36.56
Na K	1.61	1.60
Mg K	2.42	4.15
Al K	4.17	1.84
Si K	6.12	5.79
Ti K	4.12	2.94
K K	0.99	0.58
Ca K	8.97	6.82
Fe K	10.16	2.52
Ba L	17.83	2.97
Totals	100.00	

(c)

Figure 5. EDXA image of (a) synthesized of advanced ARMC shielding material. Selected area (b) EDXA outcomes (c) Elemental compositions (weight % and atomic %).

The FESEM image showed that the particles are of spherical as well as irregular in shape confirming the multiple types of chemical composition in synthesized material. The size varies in the range up to 29 - 46 nm. The EDXA spectra revealed the predominance of multi elements like Barium, titanium, Aluminum, Iron, Carbon, Magnesium, Silicon, Sodium, Calcium and Oxygen.

3.4. Permittivity and Permeability Study of Developed Advanced ARMC Shielding Material

The study of permittivity (ε) and permeability (µ) represents the dielectric and magnetic properties of the microwave absorber material. The real part of permittivity and permeability represents the storage of electric and magnetic energies respectively whereas the imaginary parts symbolize the loss and dissipation of both the energies.

The real and imaginary component of permeability for advanced ARMC shielding material with thickness, t = 1.0 mm, over the range of 1 - 13 GHz and is shown in Table 3. The table shows real and imaginary part of permeability for advanced ARMC shielding material in which the complex permeability of the sample increases with the increase in frequency till 12 GHz, after which it decreases. This can be due to natural resonance or eddy current loss. It is well known fact that when any conducting material was put in an alternating magnetic field, a close induced current will be produced inside the material, which dissipates the energy. This dissipation of energy is known as eddy current loss. Here, in the reported process as MWCNTs are used as high electrical conductivity material therefore decrease in permeability at high frequency is observed due to eddy current loss.

The value of complex permittivity including both real and imaginary part of advanced ARMC shielding material is shown in Table 3. The value shows the complex permittivity of the sample decreases with the increase in frequency. The dielectric loss may occur due to intrinsic electrical dipole moment with in the shielding material and presence of electromagnetic field due to polarization relaxation (electronic, ionic and orientation). Relatively high permittivity of the sample can cause diectric loss, mainly by dielectric polarization, spontaneous polarization and associated relaxation phenomenon. Due to the presence of MWCNTs having high electrical conductivity 1) the strong polarization, 2) dissipation of electrostatic charges and 3) ohmic losses or multiple scattering caused by the large surface area occurs, this leads to enhanced complex permittivity in the developed advanced ARMC shielding material.

3.5. EM Wave Absorption Properties of Developed Advanced ARMC Shielding Material

Reconnoiter of the EM wave absorption properties of the advanced ARMC shielding materials done by using metal plate (being anelectric conductor). A material considering for EM shielding is placed over the limit of 20 Db. The advanced ARMC material with thickness t = 1.5 mm showed a minimum R.L of −35.5 Db at 14.0 GHz with a response band width of 1.8 GHz as shown in Table 4. To obtain more R. L. value the thickness of the sample can be increased.

The unique designing and ceramic processing of chemical compositions and mineralogical phases of red mud, MWCNTs together with appropriate additive and solvent has been done. It results in the simultaneous and synergistic chemical reactions among various constituents results in the formation of tailored precursor. Further, the ceramic processing of tailored precursor powder in appropriate environment enables in formation of desired multiple ceramic phases namely barium iron titanium silicate (Bafertisite), barium aluminum silicate, iron titanium oxide, barium titanate, Barium iron titanium oxide, barium aluminium oxide, magnetite, iron sulphide (troilite) possessing multi elemental compositions and multi layered crystal structure possessing inherent properties of shielding EMI efficiently.

It is reported in literature that Barium titanate, Barium iron titanium oxides are very good as electromagnetic wave absorber material [25] . Further, in the present process the oxides of titanium and iron present in red mud reacts with barium hydroxide in the presence of alkaline ethylene alcohol leads to the formation of major EMI shielding phase namely barium titanate, Barium iron titanium and alkaline iron oxide species in the matrix of

Table 3. Complex permeability (μ) and Permittivity (ε) of advanced ARMC shielding material.

Sr. No.	Frequency (in GHz)	Complex Permeability (μ)		Complex Permittivity (ε)	
		Real part	Imaginary part	Real part	Imaginary part
1	1	0.5	0.1	4	16
2	2	0.8	0.2	3.9	14
3	3	1	0.3	3.8	13
4	4	1.2	0.4	3.4	12
5	5	1.3	0.4	2.8	11
6	6	1.4	0.45	2.5	8
7	7	1.5	0.50	2.3	6.5
8	8	1.6	0.55	2	5.8
9	9	1.65	0.7	1.8	5.3
10	10	1.7	0.8	1.75	5
11	11	1.85	1.1	1.50	4
12	12	2	2.5	1.2	3
13	13	1.3	1.8	1	2.1

Table 4. Reflection loss of advanced ARMC material.

Sr. No.	Frequency (GHz)	Reflection loss (Db)
1	0	0
2	2	0
3	4	0
4	6	−1
5	8	−6
6	10	−20
7	12	−30
8	14	−35.5
9	16	−31

in-situ functionalized MWCNTs ranging from micron to sub micron up to nanosize due to the presence of CTAB chosen as one of the surfactant. Further, the multilayer characteristics of bafertisite phase and MWCNTs imparts efficient EMI shielding to the developed advanced ARMC material.

4. CONCLUSIONS

The following conclusions can be drawn from work carried out for the novel synthesis of advanced ARMC shielding material.

1) The results obtained represent a fundamental starting point for the synthesis of advanced ARMC shielding material utilizing red mud and MWCNTs together.

2) The process enables unique designing of chemical compositions and mineralogical phases of red mud, MWCNTs with appropriate additive and solvent for obtaining in-situ synthesis of advanced EMI shielding material by ceramic processing of tailored precursor.

3) The process involves ceramic processing of tailored precursor for the formation of major EMI shielding phase namely barium titanate, bafertisite phase and alkaline iron oxide species in the matrix of in-situ functionalized MWCNTs as confirmed by XRD studies.

4) The use of CTAB chosen as one of the surfactants enables obtaining particle size of advanced ARMC material raining from micron to sub micron up to nanosize.

5) The FESEM studies revealed that the nanoparticles were almost spherical in shape with particles size in the range of 29 - 46 nm.

6) The multi layer characteristics of bafertisite phase and MWCNTs impart efficient EMI shielding to the developed ARMC.

7) The electrical conductivity of MWCNTs imparts efficient EMI shielding to the developed ARMC.

8) The presence of $BaTiO_3$ and TiO_2 imparts the EMI shielding due to high permittivity and high dielectric constant.

9) The presence of the magnetic phases due to alkaline oxides of iron increases the absorption of EMI waves.

ACKNOWLEDGEMENTS

Authors are grateful to Director CSIR-AMPRI Bhopal for providing necessary institutional facilities and encouragement.

REFERENCES

1. Mottahed, B.D. and Manoocheheri, S. (1995) A Review of Research in Materials, Modeling and Simulation, Design Factors, Testing, and Measurements Related to Electromagnetic Interference Shielding. Polymer-Plastics Technology and Engineering, 34, 271-346.

2. Neelakanta, P.S. and Subramaniam, K. (1992) Controlling the Properties of Electromagnetic Composites. Advanced Materials and Processes, 141, 20-25.

3. Lu, G.H., Li, X.T. and Jiang, H.C. (1996) Electrical and Shielding Properties of ABS Resin Filled with Nickel-Coated Carbon Fibers. Composites Science and Technology, 56, 193-200.

4. Kaynak, A., Polat, A. and Yilmazer, U. (1996) Some Microwave and Mechanical Properties of Carbon Fiber-Poly- propylene and Carbon Black-Polypropylene Composites. Materials Research Bulletin, 31, 1195-1206.

5. Melvin, G.J.H., Ni, Q.Q. and Natsuki, T. (2014) Electromagnetic Wave Absorption Properties of Barium Titanate/ Carnon Nanotube Hybrid Nanocomposites. Journal of Alloys and Compounds, 615, 84-90.

6. Micheli, D., Apollo, C., Passtrre, R. and Marchetti, M. (2010) X-Band Microwave Characterization of Carbon-Based Nanocomposite Material, Absorption Capability Comparison and RAS Design Simulation. Composites Science and Technology, 70, 400-409.

7. Qin, F. and Brosseau, C. (2012) Review and Analysis of Microwave Absorption in Polymer Composite Filled with Carbonaceous Particles. Journal of Applied Physics, 111, 061301.

8. Huo, J., Wang, L. and Yu, H.J. (2009) Polymeric Nanocomposites for electromagnetic Wave Absorption. Journal of Materials Science, 44, 3917-3927

9. Melvin, G.J.H., Ni, Q.Q., Suzuki, Y. and Natsuki, T. (2014) Microwave-Absorbing Properties of Silver Nanoparticles/ Carnon Nanotube Hybrid Nanocomposites. Journal of Materials Science, 49, 5199-5207.

10. Irwin, P., Tan, D.Q. and Cao, Y. (2010) Electromagnetic Interference Shielding Polymer Composites and Methods of Manufacture. US 7815820 B2.

11. Iyer, V.R., Ungvichian, V. and Neelakanta, P.S. (1993) A Titanium-Dioxide Based Ceramic Composite Dispersed with Conducting Inclusions as an EMI Shielding Material. 1993 IEEE International Symposium on Electromagnetic Compatibility, Dallas, 9-13 August 1993, 168-169.

12. Azadmanjiri, J., Hojati-Talem, P., Simon, G.P., Suzuki, K. and Selomulya, C. (2011) Synthesis and Electromagnetic Interference Shielding Properties of Iron Oxide/Polypyrrole Nanocomposites. Polymer Engineering & Science, 51, 247-253.

13. Ding, Z.G., Shi, S.Q., Zhang, H.L. and Cai, L.P. (2015) Electromagnetic Shielding Properties of Iron Oxide Impregnated Kenaf Bast Fiberboard. Composites Part B: Engineering, 78, 266-271.

14. Yang, J. (2006) Research on Comprehensive Utilization of Red Mud from Alumina Plant. M.S. Thesis, Huazhong University of Science and Technology, Wuhan.

15. Che, R.C., Peng, L.M., Duan, X.F., Chen, Q. and Liang, X.L. (2004) Microwave Absorption Enhancement and Complex Permittivity and Permeability of Fe Encapsulated within Carbon Nanotubes. Advanced Materials, 16, 401-405.

16. Kim, H.M., Kim, K., Lee, C.Y., Joo, J., Cho, S.J., Yoon, H.S., et al. (2004) Electrical Conductivity and Electromagnetic Interference Shielding of Multiwalled Carbon Nanotube Composites Containing Fe Catalyst. Applied Physics Letters, 84, 589-591.

17. Xiang, C.S., Pan, Y.B., Liu, X.J., Sun, X.W., Shi, X.M. and Guo, J.K. (2005) Microwave Attenuation of Multiwalled Carbon Nanotube-Fused Silica Composites. Applied Physics Letters, 87, 123103.

18. Yang, Y.L. and Gupta, M.C. (2005) Novel Carbon Nanotube-Polystyrene Foam Composites for Electromagnetic Interference Shielding. Nano Letters, 5, 2131-2134.

19. Huang, Y., Li, N., Ma, Y.F., Du, F., Li, F.F., He, X.B., Lin, X., Gao, H.J. and Chen, Y.S. (2007) The Influence of Single-Walled Carbon Nanotube Structure on the Electromagnetic Interference Shielding Efficiency of Its Epoxy Composites. Carbon, 45, 1614-1621.

20. Verma, S. and Amritphalae, S.S. (2015) A Novel Process for Preparation of Hybrid Nanosized Neodymium Oxybromide and Neodymium Oxide. Russian Journal of General Chemistry, 85, 1528-1531.

21. Chen, W., Mai, L.Q., Qi, Y.Y. and Dai, Y. (2006) One-Dimensional Nanomaterials of Vanadium and Molybdenum Oxides. Journal of Physics and Chemistry of Solids, 67, 896-902.

22. (1984) Powder Diffraction File, Alphabetical Index Inorganic Phases. Published by JCPDS International Centre for Diffraction Data 1601, Park Lane Swarthmore, Pennsylvania.

23. Rahman, M.N. (2003) Ceramic Processing and Sintering. CRC Press, Taylor and Francis Group, London.

24. Zhang, M.-S., Yin, Z., Chen, Q., Zhang, W.F. and Chen, W.C. (2001) Study of Structural and Photoluminescence Properties in Barium Titanate Nanocrystals Synthesis by Hydrothermal Process. Solid State Communications, 119, 659-663.

25. Sukanta, D., Nayak, G.C., Sahu, S.K., Routray, P.C., Roy, A.K. and Baskey, H. (2014) Microwave Absorption Properties of Double-Layer RADAR Absorbing Materials Based on Doped Barium Hexaferrite/TiO2/Conducting Carbon Black. Journal of Engineering, 2014, Article ID: 468313.

8

ENVIRONMENTAL EFFECTS ON THE CONSTRUCTION AND PHYSICAL PROPERTIES OF BOMBYX MORI COCOONS

C. Offord[1,2], F. Vollrath[1], and C. Holland[1,3]

[1]Department of Zoology, University of Oxford, South Parks Road, Oxford OX1 3PS, UK

[2]Department of Ecology and Evolutionary Biology, Princeton University, Princeton, NJ 08544-2016, USA

[3]Department of Materials Science and Engineering, University of Sheffield, Mappin Street, Sheffield S1 3JD, UK

ABSTRACT

Published studies of silks focus on processed fibres or the optimum conditions for their production. Consequently, the effects of the environment on the physical properties of the cocoon are either poorly understood or kept as closely guarded industrial secrets. In this study, we test the hypothesis that silkworms as ectothermic animals respond to environmental conditions by modifying their spinning behaviour in a predictable manner, which affects the material properties of the cocoons in predictable ways. Our experiments subjected spinning*Bombyx mori* silkworms to a range of temperatures and relative humidities that, as we show, affect the morphology and mechanical properties of the cocoon. Specifically, temperature affects cocoon morphology as well as its stiffness and strength, which we attribute to altered spinning behaviour and sericin curing time. Relative humidity affects cocoon colouration, perhaps due to tanning agents. Finally, the water content of a cocoon modifies sericin distribution and stiffness without changing toughness. Our results demonstrate environmentally induced quality parameters that must not be ignored when analysing and deploying silk cocoons, silk filaments or silk-derived bio-polymers

INTRODUCTION

Silk production is a feature unique to the arthropods with approximately 98 % of Lepidoptera species producing some form of silk [1]. Silks are formed inside but perform outside the body, requiring specialisations to both internal processing and external climatic impact [2]. As silkworms are ectothermic (i.e. follow the outside temperature), the environment during spinning plays a crucial role in both physiology and behaviour, with consequences for the properties and performance of the silks.

The larvae of the domesticated Chinese mulberry silkmoth, *Bombyx mori* construct a protective cocoon before undergoing metamorphosis. Spinning takes place over several hours as the larva draws a continuous silk thread from labial glands in its head in a figure-of-eight fashion [3]. As a material, the cocoon is a random-fibre non-woven composite [4, 5]. It consists of two principle proteins, heavy and light chain fibroin (which form the fibre) and at least five sericin 'gum' proteins (which coat and bind the fibres together and account for 20–30 % of the cocoon mass) [6–8]. Environmental conditions contribute towards the properties of the spun fibre through the silkworm's behaviour (i.e. processing) as well as through direct interaction with the material itself. At higher ambient temperatures the rate of spinning is faster than at lower ones [9]; and at very low temperatures silkworms fail altogether to spin cocoons [10]. Spinning behaviour has been studied in some detail within individuals [11] and between strains [3] and it may be manipulated internally by hormone treatment [10], by the administration of drugs, or by surgical procedures [12]. When tested individually, silk fibres show considerable variation on intraspecific and interspecific scales and the rate of spinning has been shown to affect fibre structural and mechanical properties [13–17] and in particular the effect of degumming prior to use [18]. Beyond individual fibres, mechanical tests on the cocoon have been performed between cocoons produced by an individual silkworm [19], across different dimensions of the cocoon [15] and between individual layers of the cocoon [15, 20, 21].

This study aims to determine the effects of environment on spinning by measuring the effects of relative humidity (RH) and temperature on the physical properties of the cocoon of *B. mori*. Ninety silkworms, divided into nine groups, constructed cocoons under three temperature and three humidity conditions in a factorial design. Temperature and humidity treatments were chosen to represent a spread around optimum conditions, taken as 25 °C and an intermediate relative humidity (RH) of approximately 65–75 % [8, 19, 22]. We assert that this bio-centric approach can provide valuable and novel insight into Nature's ways to control, as well as constrain, its composite materials.

MATERIALS AND METHODS

A single batch rearing of *B. mori* silkworms derived from a 4 way poly-hybrid cross of 2 Japanese and 2 Chinese strains in the fifth, final instar and raised on mulberry leaves in controlled conditions in the laboratory (23 ± 2 °C 60–70 % RH) were used for testing. Silkworms about to start spinning were placed in individual, pre-weighed 50 ml plastic tubes with holes drilled in the lid to allow for the circulation of air. Nine Perspex containers with a volume of 3.75 l were filled with ten tubes each, arranged in two rows, and sealed to maintain a constant RH independent of the outside atmosphere.

Each Perspex container (and thus the silkworm within) was randomly assigned a combination of temperature and humidity conditions. RH was controlled by placing a wide plastic tub alongside the falcon tubes. In containers assigned a low RH (10 %), the tubs were filled with desiccated silica gel. In containers assigned a high RH (100 %), the tubs were filled with 200 ml of Type 1 (ultra-purified) water. Those assigned an intermediate humidity contained tubs filled with 200 ml of saturated glucose solution, demonstrated to maintain a constant RH across the temperature range used, 15–35 °C [23]. Conditions were monitored using a temperature and humidity monitor (HOBO instruments UK), and the glucose solution was shown to maintain an intermediate relative humidity of 85 % (±5 % error), close to standard spinning conditions. Each box was then sealed and temperature controlled in either incubators (15 and 35 °C) or air conditioning (25 °C). All boxes were stored in the dark for consistency, and left undisturbed for five days to allow the silkworms to spin cocoons.

After five days, the tubes were weighed and photographed. The length and diameter of the cocoons were recorded using a digital vernier calliper (accuracy ± 0.01 mm). The cocoons were cut open circumferentially with a razor blade and the silkworms removed. The state of the silkworms was noted, as not all silkworms had begun to pupate. Cocoons were then divided into sections to provide experimental samples (see Fig. S3 in supplementary materials). A hole punch was used to punch two 6-mm discs, 5 mm apart, from each half, with the section in-between serving as the 'bone' for tensile testing. Separate discs were used to obtain mass, colouration, number of layers and water content. Tweezers were used to peel layers of the disc apart under a Nikon dissection microscope and the number of layers was recorded. For sericin distribution analysis sections from intermediate layers of cocoons were sputter coated (SC7620, Quorum Technologies Ltd) for 300 s at 18 mA with an 80:20 gold palladium target producing a coating of approximately 25 nm, then imaged using a scanning electron microscope (SEM) (Neoscope 2000, Nikon Instruments UK) at ×100 and ×200 magnification at 10 keV.

Mechanical testing was conducted using an Instron Tensile Tester with a 500 N load cell (model 5512, Instron, UK) on at least two samples per cocoon. The thickness and width of each 'bone' segment was measured using a vernier calliper to determine the area of the cross-section of the sample at its narrowest point and was assumed to be rectangular for stress calculations. The humidity and temperature of the room at time of testing was also recorded. Samples were clamped at a gauge distance of 6 mm and the clamps moved apart at a rate of 2.4 mm/min, and hence at a strain rate of 0.067 s^{-1}. The cue to end the test was prompted when the load had dropped by 90 %. Video of the silk samples' breaking was captured using a JVC ccd camera and a customised Matlab program (courtesy of A. Baker). These files were used to ensure that samples had broken correctly (i.e. at the centre of the sample, rather than at the clamps— Fig. S4 in supplementary materials) and exclude defective samples from analysis. Audio files were recorded by a modified computer microphone with Audacity version 1.2.6. These audio files were used in further analysis to show the breaking of silk fibres in the sample, a method not previously described in this area. Young's modulus was calculated as the gradient of the curve from 2–5 % strain, instead of the standard 0–0.2 % strain [24], to minimise the effects of any initial curvature of the sample due to the surface curvature of the cocoon.

There was considerable variation in cocoon coloration. To quantify colour differences among individual cocoons, a white grid with a black background was printed, onto which one disc from each cocoon was positioned. The grid was lit by fibre lamps and photographed with a Canon A640 digital camera. To analyse the colour differences across the grid, ImageJ (http://rsbweb.nih.gov/ij/) was used to obtain RGB histograms of 5-mm-diameter circular cross-sections of each photographed disc. Of the RGB histograms, the blue channel was used as an indicator of the level of discolouration of the silk and used to calculate a standard measure of tanning by subtracting the blue value of each sample from the blue value of the least discoloured (i.e. palest) of all the samples.

To quantify water content, discs from 37 cocoons, chosen randomly from the sample set to represent each group, were used for Thermo-Gravimetric Analysis (TGA, Q50, TA Instruments UK). The sample mass was continually recorded, whilst discs were individually heated in pans for 80 min to 150 °C.

The effects of temperature and humidity on the physical properties of the cocoon were simultaneously analysed using general linear models in Minitab 15.1. p-values for significant factors (boundary for significance: $p = 0.05$) calculated from ANOVA (analysis of variance) are displayed where appropriate.

RESULTS

Overall, the proportion of silkworms surviving each temperature treatment was consistent with recommended conditions for sericulture [9, 22]. The highest proportion of larvae survived to produce cocoons at 25 °C, with a notable exception that within this group mortality was highest at 85 % RH at 25 °C, which we are currently unable to account for (see Fig. S1 in supplementary materials). However, taken as a whole and considering the extremes of our test environments, the group exposed to 100 % RH and 35 °C had the lowest survival score yet 90 % of the larvae survived and successfully produced cocoons at 10 % RH and 15 °C, suggesting that silkworms are less tolerant of high RH and temperature treatments during spinning.

The physical properties of the cocoons varied across the treatment combinations. Silkworms produced longer, thinner cocoons at higher temperatures ($p < 0.0005$) (Fig. 1). With the increasing RH, the ease with which fibres of the disc cut from the cocoon could be peeled apart increased, but the distinction between individual layers decreased. Whilst there was no significant variation in cocoon thickness

($p = 0.163$, $p = 0.719$ for temperature and RH, respectively), the mass of discs cut from the cocoon varied significantly with RH ($p < 0.0005$) and temperature ($p = 0.003$), with the greatest disc mass occurring at the intermediate temperatures (Fig. S2 in supplementary material).

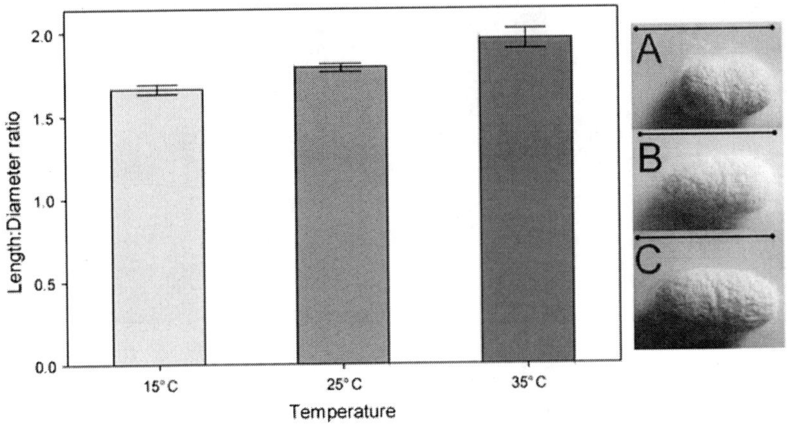

Figure 1. Environmental effects on morphological properties of cocoons. Effects of temperature on shape with length:diameter ratios all significantly different at $p < 0.0005$. ($n = 69$). *Error bars* show ±1 SEM. Photographs are representative cocoons spun at 10 % RH and *A* 15 °C, *B* 25 °C and *C* 35 °C. *Scale bar* is 4 cm

SEM micrographs show cocoon samples from both extreme RH treatments at the control temperature of 25 °C (Fig. 2). There are two striking differences between the structures of the two cocoons. Firstly, individual fibres produced at low humidity are clearly separate threads whereas at high humidity the distinctiveness seems to disappear. We observe that at high RH the sericin coating of the fibres has spread and thus immersed multiple fibres and diffused the cocoon layer. The failure mechanisms of the cocoons as a result of humidity can be qualitatively determined by the degree of sericin fragmentation as layers were peeled apart in order to prepare samples for SEM. At low RH layer separation resulted in fragmentation at the fibre sericin interface, suggesting a more brittle material whereas at higher RH these deposits were not visible and the fibres had a smoother appearance. This observation of material differences is supported by the use of sound recordings during tensile testing (insets (i) and (ii) in Fig. 2). Samples from cocoons spun in the lowest RH treatment show a single breaking of the fibre attachments to the sericin matrix fibres, as indicated by a single spike in the sonogram, whereas multiple breaks are recorded from cocoons spun at high RH. This is all consistent with the qualitative observations made during layer peel tests, with lower RH samples being harder to pull apart but the distinction between layers more prominent.

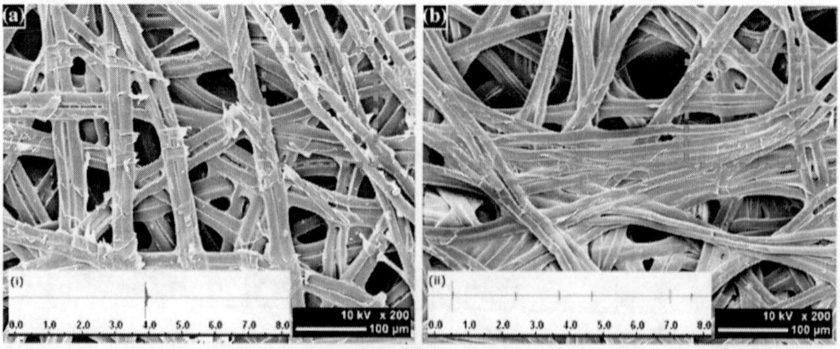

Figure 2. SEM micrographs of intermediate layers of two cocoons produced at 25 °C under 10 % (**a**) and 100 % RH (**b**). *Insets* (*i*) and (*ii*) show sound files for silk layers breaking over a period of 8 s under 10 and 100 % RH, respectively. Magenta rectangles highlight areas of the SEM which clearly demonstrate the difference in sericin morphology (fragmentation and distribution between fibres) as a result of humidity

In order to confirm whether these observations impacted the mechanical performance of the cocoon as a whole, samples were subjected to tensile testing. We observed that silkworm cocoons under tension were affected significantly by temperature and humidity conditions during spinning (Fig. 3). Stress–strain curves for cocoons spun at low RH showed a sudden, single break, compared to curves for cocoons spun at high RH, which displayed a more gradual decrease in stress once the material had reached its maximum stress.

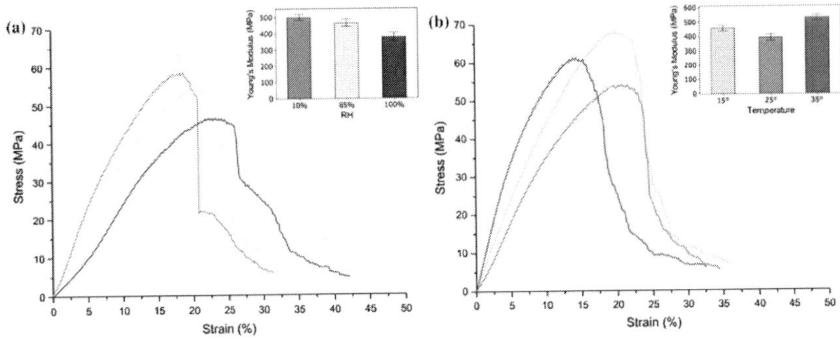

Figure 3. Environmental effects on the tensile properties of cocoons when spun under different conditions and tested under the same (25 °C 50 % RH). **a** Tensile behaviour of representative cocoon samples taken from the three different humidity treatments: 10 % RH *light green line*, 85 % RH *light blue line*, 100 % RH *navy line* (all at 25 °C). **b** Curves for samples of cocoons taken from three different temperature treatments: 15 °C *yellow line*, 25 °C *orange line*, 35 °C *red line* (all at 85 % RH). *Insets* show averaged modulus data for humidity and temperature. Effects of both spinning conditions significant at $p < 0.0005$ for both temperature and RH ($n = 62$). *Error bars* show ±1 SEM

Young's modulus (stiffness), the breaking energy (toughness), the strain at maximum stress and maximum strain all showed variation across the treatments. Young's modulus varied significantly with temperature and RH ($p < 0.0005$ for both) (Fig. 3, Table 1). Interestingly, the lowest modulus was observed in the cocoons spun at the intermediate temperature. Stiffer cocoons were produced at 15 °C and the stiffest at 35 °C. Cocoons spun in higher humidity treatments had lower average moduli, and Young's moduli showed a strong negative correlation with the mass of the disc cut from a cocoon (Pearson's correlation coefficient: −0.547, $p < 0.0005$).

The average breaking energy (toughness) for the cocoon samples was lower for cocoons spun at higher temperatures ($p < 0.0005$) (Table 2). However, breaking energy remained constant across humidity treatments. This might be attributed to a trade-off between stiffness and degree deformation whereby increasing the humidity decreased the stiffness but increased the maximum strain resulting in a consistent toughness in the experimental RH range (Fig. 3a).

Table 1. Young's modulus (GPa) across temperature and RH conditions during spinning

Young's modulus	Temperature			
	15 °C	25 °C	35 °C	Average
RH				
10 %	0.50 (0.03)	0.46 (0.03)	0.57 (0.03)	0.50 (0.02)
85 %	0.44 (0.05)	0.42 (0.03)	0.53 (0.02)	0.47 (0.02)
100 %	0.42 (0.03)	0.30 (0.04)	0.47 (0.05)	0.38 (0.03)
Average	0.46 (0.02)	0.40 (0.02)	0.53 (0.02)	

SEM shown in brackets

Effect of temperature $p < 0.0005$

Effect of RH $p < 0.0005$ ($n = 62$)

Table 2. Breaking energy (MJkg^{-1}) across temperature and RH conditions during spinning

Breaking energy	Temperature			
	15 °C	25 °C	35 °C	Average
RH				
10 %	1.4 (0.1)	1.2 (0.1)	0.8 (0.1)	1.2 (0.1)
85 %	1.3 (0.1)	1.2 (0.1)	1.1 (0.1)	1.2 (0.1)
100 %	1.4 (0.2)	1.1 (0.1)	1.2 (0.1)	1.2 (0.1)
Average	1.4 (0.1)	1.2 (0.1)	1.0 (0.1)	

SEM shown in brackets

Effect of temperature $p < 0.0005$

Effect of RH $p = 0.996$ ($n = 62$)

Considerable variation in the colour of the cocoons was observed from the creamy white colour typical of *B. mori* cocoons to a darker, tanned colour (Fig. 4). Humidity had a strong effect on the average discolouration of the cocoon disc ($p < 0.0005$), although there was no significant effect of temperature. The discolouration was most evident in cocoons spun under the highest RH conditions.

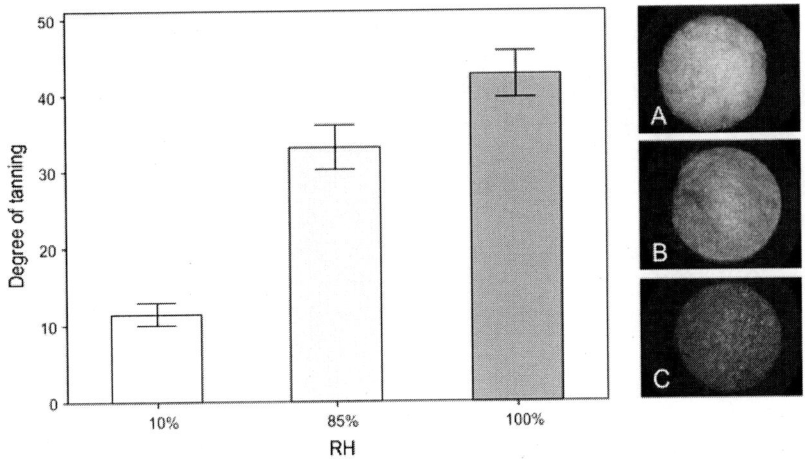

Figure 4. Effects of humidity on cocoon colouration (tanning). Degree of tanning is measured on an arbitrary scale. $p < 0.0005$ ($n = 69$). *Error bars* show ±1 SEM. Photographs *A-C* show discs of 6 mm in diameter taken from cocoons produced by *B. mori* spinning at 10 % RH, 15 °C (*A*); at 100 % RH, 35 °C (*B*); and the wild silkmoth larva, *Atticas atlas*, (*C*). Photos taken under controlled lighting conditions

Table 3. Environmental effects during spinning on percentage water content of cocoon

% Mass water	Temperature			
	15 °C	25 °C	35 °C	Average
RH				
10 %	25 (5)	16 (1)	13 (7)	18 (3)
85 %	41 (2)	42 (1)	29 (6)	36 (3)
100 %	41 (6)	47 (5)	45 (2)	44 (2)
Average	36 (3)	35 (4)	31 (4)	

SEM shown in brackets

Effect of temperature $p < 0.0005$

Effect of RH $p < 0.0005$ ($n = 37$)

Due to the significant influence of RH on morphological and mechanical properties of cocoons, thermogravimentric analysis was used to determine the percentage water content by taking the mass of discs at 21 °C and then again after heating to 150 °C ($n = 37$, Table 3). Water content was shown to be highly dependent on humidity across all treatments ($p < 0.0005$), and at 150 °C, the masses of all discs were largely invariant (temperature $p = 0.07$ and humidity$p = 0.801$). This indicates that water was responsible for the difference in mass and the amount of silk protein used in cocoon construction in response to different environmental conditions does not change.

DISCUSSION

Our data clearly demonstrate how key environmental conditions during spinning affect both the morphological and the mechanical properties of *B. mori* cocoons.

The influence of 'climate' on cocoon morphology can be separated into two categories; overall shape (affected by temperature) and cocoon colour (affected by humidity). Larvae produced longer, thinner cocoons at higher temperatures (Fig. 1). Given that the silkworm spins from its head yet is attached to the substrate by its abdominal pro-legs, there is a direct causal link between body length and the shape of the cocoon [3]. At higher temperatures the silkworms may be able to extend their bodies further, leading to longer cocoons, which may in turn be used to regulate heat dissipation. A long, thin cocoon has a higher surface area to volume ratio than a short, round one, hence an ability to vary cocoon shape as a response to different temperature conditions during spinning could help maintain a suitable temperature range (and by extension also different internal vapour pressures [25]) for metamorphosis. The effects of humidity on the cocoon were visibly more pronounced. Higher humidities during spinning led to a gradual discolouration of cocoons from white to a golden or light brown colour after several days (Fig. 4). Silkworm faeces or bacterial infection are highly unlikely to have been responsible for this discolouration considering the even distribution of colour across the surface of the cocoon and the antibacterial properties of the sericin protein [26]. A possible cause of discolouration would be the presence of tanning agents as the colour is akin to that found in some wild silkmoth cocoons (e.g. *Atticas atlas*). Brunet and Coles reported that *Antheraea pernyi* cocoons showed discoloration from white to brown under high humidity conditions. They reported that moisture in the silk enables enzymatic cleavage and oxidation of endogenous tanning phenols which in turn both changes the cocoon's colour and crosslinks protein chains, the latter having a strong effect on the mechanical properties of single silk fibres [27]. Whilst previous studies have used *B. mori* as a control for the presence of tanning [28] our results indicate that through careful modification of spinning conditions it may be possible to determine the agents and effects this discolouration has on the silkworm cocoon. These observations are of particular importance to agriculture as the quality and grading of cocoons is largely dependent on both size and purity [29–31].

Humidity and temperature had differing and combinative effects on the mechanical properties of the cocoon (Fig. 2). Broadly speaking, temperature modified the stiffness and strength of the cocoon through either increased spinning rate (35 °C) or sericin curing time (15 °C). Humidity altered the stiffness (pliability) and the way in which a cocoon broke, but did not alter the overall toughness. Silkworms have been shown to increase their rate of spinning at higher temperatures [9] and forced-spun silkworms show an increase in fibre strength and stiffness with increased reeling rate [17, 32]. Analysis of our SEM images of cocoons spun at different temperatures shows a negative correlation between fibre thickness and temperature, with fibres spun at 35 °C being at least 25 % thinner than those spun at 15 °C, consistent with observations of diameter decrease with increased spinning rate in silkworms [32] and spiders [33, 34]. Therefore it is most likely that the fibre properties in cocoons spun at higher temperatures explain the higher stiffness and strength than those spun at 25 °C. A means to test this hypothesis would be to degum, unravel and subsequently test single fibre properties. This is an area of ongoing study, although recent work has highlighted potentially confounding influences of the degumming process in such an assessment [18, 35]. However, surprisingly at lower temperatures, there is also an observed increase in strength and stiffness of the cocoon upon comparison to 25 °C. This we believe could be attributed to the role of the other constituent of the cocoon, the sericin. In order for the cocoon to have any form at all, it is crucial that the fibres are adequately bound together, which occurs as the sericin coating of the fibres dries/cures. We propose that at lower temperatures the sericin takes longer to cure and as such forms stronger connections between fibres, leading to a greater overall structural integrity, a hypothesis consistent with our current understanding of cocoon mechanical properties [4, 5,36].

The role of humidity during construction of the cocoon is also readily accounted for. Our data indicate that the water content of the cocoon is dependent upon the environmental conditions at point of spinning and not the subsequent storage environment (as all samples were kept in the same environment post cocoon construction, Table 3). Therefore the significant differences in hydration reported are either the result of the retention of solvent water during cocoon construction, or a tendency for a specific cocoon type to absorb more atmospheric water during storage than another [25]. Regardless of the source of the water its influence may be seen in the morphological differences in the distribution of sericin (Fig. 2) and the definition of layers in the cocoon. This observation suggests that methods aiming to distinguish individual cocoon 'layers', may not be transferrable across a wide range of environmental conditions [15, 20, 21]. An explanation for the differences in mechanical properties lies within the sericin. Sericin readily absorbs water [26], and the increased water content of the binder acts as a plasticiser and leads to a reduced stiffness and strength of the cocoon, as evidenced by our tensile data (Fig. 3), the reduction in sericin fragmentation (Fig. 2) and the ease at which individual fibres could be pulled from the cocoon. Therefore the distinction of layers may be dependent on the environmental drying time and the speed of the animal as it puts down additional layers. Not only does trapped water alter the stiffness and strength of the cocoon but also the mechanism by which it fails, a novel observation revealed here through the introduction of sonic analysis (Fig. 2). However, despite these changes in the properties of the cocoon, the breaking energy (toughness) remains constant across humidities. The toughness of a material corresponds to its capacity for energy dissipation and is therefore the most biologically relevant mechanical property of a cocoon as it determines the resilience of the cocoon to the environment [37]. Given that *B. mori* cocoons show mechanical behaviour comparable to wild silkmoth cocoons [4] it seems that the mechanical toughness of the cocoon could still be relevant in *B. mori,*despite a long history of domestication in this species. To place this into a biological context, during the periods of the year in which silkworms naturally spin their cocoons, it is far more likely to have environmental variations in seasonal humidity than temperature. Hence a greater tolerance for humidity in maintaining the toughness of the cocoon is crucial to enable the cocoon to withstand attacks from predators or parasitoids whilst providing an optimum internal environment for pupation [25].

This study explored the crucial relationship between the environmental spinning conditions of silkworms and the properties of the cocoons produced. Combining classical biological experiments with analytical techniques from engineering and material sciences allows us to better comprehend the selection pressures on the cocoon phenotype. The novel observation of environmentally determined 'tanning' in *B. mori* cocoons opens the field for study into potential causes, and their apparent suppression in commercial silk production. Furthermore, the results of tensile testing reveal the dependence of behaviour and mechanical properties on temperature and to some extent their tolerance of humidity in spite of the huge variations in water content of the cocoon. Evidently the environmental conditions during spinning play an important role in the silkworm's construction of the cocoon and must be integrated into future assessment of the sources of variation in experiments with silk, from single fibre tests to reconstituted feedstock production [38].

ACKNOWLEDGEMENTS

We thank Tom Gheysens, Fujia Chen and David Porter for assistance with mechanical testing and discussion. We thank the Department of Zoology and Magdalen College, Oxford University, the US-AFOSR (FA9550-12-1-0294), the European Research Council (SP2-GA-2008-233409) and the EPSRC (EP/K005693/1).

CONFLICTS OF INTEREST

The authors declare they have no conflicts of interest.

REFERENCES

1. Craig C (1997) Evolution of arthropod silks. Ann Rev Entomol 42:231–267

2. Vollrath F, Porter D, Holland C (2013) The science of silks. MRS Bull 38(01):73–80.

3. Kiyosawa M, Ito E, Shirai K, Kanekatsu R, Miura M, Kiguchi K (1999) Cocoon spinning behavior in the silkworm, *bombyx mori*: comparison of three strains constructing different cocoons in shape. Zoolog Sci 16:215–223

4. Chen F, Porter D, Vollrath F (2012) Structure and physical properties of silkworm cocoons. J R Soc Interface.

5. Chen F, Porter D, Vollrath F (2012) Silk cocoon (*Bombyx mori*): multi-layer structure and mechanical properties. Acta Biomater 8(7):2620–2627

6. Sprague KU (1975) The *Bombyx mori* silk proteins: characterization of large polypeptides. Biochemistry 14(5):925–931

7. Suzuki Y (1977) Differentiation of silk gland. A model system for the study of differential gene action. In: Beermann W (ed) Results and problems in cell differentiation. Springer, Heidelburg, pp 1–44

8. Sehnal F (2008) Prospects of the practical use of silk sericins. Entomol Res 38:S1–S8.

9. Ramachandra YL, Bali G, Rai SP (2001) Effect of temperature and relative humidity on spinning behaviour of silkworm (*Bombyx mori.*L). Indian J Exp Biol 39(1):87–89

10. Sehnal F, Akai H (1990) Insect silk glands: their types, development and function, and effects of environmental factors and morphogenetic hormones on them. Int J Insect Morphol Embryol 19(2):79–132

11. Miura M, Hoshi H, Nishioka T (1990) Analysis of the spinning behaviour of silkworms. J Seric Sci Jpn 59:118–126

12. Wg VDK, Cm W (1954) Cocoon construction by the Cecropia silkworm: III. the alteration of spinning behavior by chemical and surgical techniques. Behaviour 6:233–255

13. Perez-Rigueiro J, Elices M, Llorca J, Viney C (2001) Tensile properties of silkworm silk obtained by forced silking. J Appl Polym Sci 82(8):1928–1935

14. Reed EJ, Bianchini LL, Viney C (2012) Sample selection, preparation methods, and the apparent tensile properties of silkworm (*B. mori*) cocoon silk. Biopolymers 97(6):397–407.

15. Zhao HP, Feng XQ, Yu SW, Cui WZ, Zou FZ (2005) Mechanical properties of silkworm cocoons. Polymer 46(21):9192–9201

16. Shao ZZ, Vollrath F (2002) Surprising strength of silkworm silk. Nature 418(6899):741–741

17. Mortimer B, Holland C, Vollrath F (2013) The forced reeling of Bombyx mori silk: separating behaviour and processing conditions. Biomacromolecules 14(10):3653–3659

18. Perea GB, Solanas C, Marí-Buyé N, Madurga R, Agulló-Rueda F, Muinelo A, Riekel C, Burghammer M, Jorge I, Vázquez J, Plaza GR, Torres AL, del Pozo F, Guinea GV, Elices M, Cenis JL, Pérez-Rigueiro J (2016) The apparent variability of silkworm (Bombyx mori) silk and its relationship with degumming. Eur Polymer J 78:129–140.

19. Huang SQ, Zhao HP, Feng XQ, Cui W, Lin Z, Xu MQ (2008) Mechanical properties of cocoons constructed consecutively by a single silkworm caterpillar, *Bombyx mori*. Acta Mechanica Sinica 24(2):151–160.

20. Zhao HP, Feng XQ, Cui WZ, Zou FZ (2007) Mechanical properties of silkworm cocoon pelades. Eng Fract Mech 74(12):1953–1962

21. Zhang J, Kaur J, Rajkhowa R, Li JL, Liu XY, Wang XG (2013) Mechanical properties and structure of silkworm cocoons: a comparative study of *Bombyx mori*, Antheraea assamensis, Antheraea pernyi and Antheraea mylitta silkworm cocoons. Mater Sci Eng C 33(6):3206–3213.

22. Aruga H (1994) Principles of sericulture. CRC Press

23. Winston PW, Bates DH (1960) Saturated solutions for the control of humidity in biological research. Ecology 41:232–237.

24. Shackelford JF (2005) Introduction to materials science for engineers, 6th edn. Pearson Education, Inc., Upper Saddle River, NJ

25. Horrocks NPC, Vollrath F, Dicko C (2013) The silkmoth cocoon as humidity trap and waterproof barrier. Comp Biochem Physiol A 164(4):645–652.

26. Zhang YQ (2002) Applications of natural silk protein sericin in biomaterials. Biotechnol Adv 20(2):91–100

27. Brunet PCJ, Coles BC (1974) Tanned silks. Proc R Soc London Ser B-Biol Sci 187(1087):133–170

28. Raven DJ, Earland C, Little M (1971) Occurrence of dityrosine in Tussah silk fibroin and keratin. Biochimica et Biophysica Acta (BBA)—Protein. Structure 251(1):96–99.

29. Sonwalkar TN (1993) Handbook of silk technology, 1st edn. New Age International, New Delhi

30. Gu G, Jiang L (2002) Changes in the grading standards and prices of Chinese silkworm cocoons in China. Silk 11:44–47

31. Guan J, Porter D, Vollrath F (2013) Thermally induced changes in dynamic mechanical properties of native silks. Biomacromolecules 14(3):930–937.

32. Mortimer B, Guan J, Holland C, Porter D, Vollrath F (2015) Linking naturally and unnaturally spun silks through the forced reeling of *Bombyx mori*. Acta Biomater 11:247–255.

33. Madsen B, Shao ZZ, Vollrath F (1999) Variability in the mechanical properties of spider silks on three levels: interspecific, intraspecific and intraindividual. Int J Biol Macromol 24(2–3):301–306

34. Vollrath F, Madsen B, Shao Z (2001) The effect of spinning conditions on the mechanics of a spider's dragline silk. Proc R Soc Biol Sci 268(1483):2339–2346

35. Jiang P, Liu H, Wang C, Wu L, Huang J, Guo C (2006) Tensile behavior and morphology of differently degummed silkworm (Bombyx mori) cocoon silk fibres. Mater Lett 60(7):919–925.

36. Chen F, Porter D, Vollrath F (2010) Silkworm cocoons inspire models for random fiber and particulate composites. Phys Rev E 82(4):041911

37. Porter D, Vollrath F (2009) Silk as a biomimetic ideal for structural polymers. Adv Mater 21(4):487–492

38. Boulet-Audet M, Holland C, Gheysens T, Vollrath F (2016) Dry-spun silk produces native-like fibroin solutions. Biomacromolecules.

9

INFLUENCE OF INERTIA AND LOW ACTIVE MINERAL ADMIXTURE ON STRENGTH AND MICROSTRUCTURE OF CEMENT-BASED MATERIALS

Meijuan Rao,[1,2] Wan Tang,[3] Wei Zhou,[3] Yaning Kong,[3] and Shuhua Liu[3]

[1]Changjiang River Scientific Research Institute, Wuhan 430010, China
[2]State Key Laboratory of Silicate Materials for Architectures, Wuhan University of Technology, Wuhan 430070, China
[3]State Key Laboratory of Water Resources and Hydropower Engineering Science, Wuhan University, Wuhan 430072, China

ABSTRACT

Cement-based materials were investigated by comparing the strength and microstructure of pastes and mortar containing limestone powder or low quality fly ash. The compressive strength of the mortar at 28 and 90 d was examined whose microstructures were characterized by X-ray diffraction (XRD), scanning electron microscopy (SEM), thermogravimetric analysis, and differential thermal analysis (TG-DTA). The results indicated that the strength of mortar decreased with increasing mineral admixtures. The limestone powder mainly acted as inert filler and hardly took part in the chemical reaction. Low quality fly ash may accelerate the formation of hydration products in samples with more chemically bonded water. This further resulted in a higher degree of cement hydration and denser microstructure, while the overall heat of hydration was reduced. At the early stage of hydration, low quality fly ash can be considered as an inert material whereas its reactivity at the later stage became high, especially for ground low quality fly ash.

1. INTRODUCTION

Cement is widely used in the production of concrete with its annual consumption close to 2 billion tons [1–3]. The production of cement itself however is quite energy-intensive and responsible for large quantities of CO_2emissions. In addition, the high amount of alkali in cement might cause cracks in the structure of concrete [4]. As one of the solutions to these, nowadays many mineral admixtures have been used as supplementary cementitious materials (SCMs) [5, 6] to reduce the amount of cement clinker, with other excellent properties such as reduced heat of hydration and improved durability of concrete.

Mineral admixtures can be divided into cementitious materials, pozzolanic active materials, and inert materials according to their chemical activities [7]. Some studies have shown that the inert mineral admixture can enhance the properties of concrete in terms of strength, frost resistance, and so forth [8–11]. This is mainly due to the dilution effect of the inert mineral admixture on cement [11]. It has been reported that equivalent amount of low active or inert mineral admixture exhibited no difference in the early hydration of cementitious materials [12, 13]. Recently, limestone powder has been applied widely in construction engineering as a new type of concrete admixture [14–16]. The particle size distribution and fineness of limestone powder are similar to those of fly ash, both of which have physical activity, morphological effect, and filling effect. The use of limestone powder as admixture in concrete leads to a better accumulation effect in cement particle skeleton and offers crystalline nucleus to the formation of calcium hydroxide (CH) and calcium silicate hydrate (C-S-H) gel. These synergistic effects can accelerate the process of cement hydration. It is traditionally believed that limestone powder is an inert admixture with high fineness which may be applied in concrete to make up composite cementitious material system that gives full play to its microaggregate effect [17–19]. Fly ash is used in concrete in the building material industry, construction engineering, and road engineering accounts for about 90% of its overall use [20, 21]. Although the progress in fly ash utilization is at the forefront compared to some developed countries, the utilization level is still very low. Research in fly ash is mainly focused on its physical and chemical properties, but few studies have reported low quality fly ash. In order to promote further use of fly ash, it is therefore critical to say something here about the significance of this work.

In this paper, a comparative system was for the first time established to assess the impacts of limestone powder as an inertia mineral admixture and fly ash as a low active mineral admixture on the strength and microstructure of cement-based materials. This is of great significance in improving the level of low quality fly ash utilization, thereby conserving resources and protecting the environment.

2. EXPERIMENTAL

2.1. Raw Materials

Ordinary Portland cement P.O 42.5 is supplied by Huaxin Cement Plant. Low quality fly ash is supplied by Guanyinyan Hydropower Project. Table 1 shows the chemical compositions of the cement, fly ash, and limestone powder determined by X-ray Fluorescence (XRF).

Table 1. Chemical composition of cement, low quality fly ash, and limestone powder/mass, %.

Compositions	SiO_2	Fe_2O_3	Al_2O_3	CaO	MgO	K_2O	Na_2O	SO_3	LOI
Cement	23.61	3.44	4.06	59.59	2.51	0.86	0.19	—	3.00
Low quality FA	53.41	4.28	25.79	2.60	2.79	1.38	0.42	0.3	3.67
Limestone powder	1.79	0.35	0.56	54.69	0.40	0.14	—	0.03	41.93

Figure 1 is the particle morphology of raw ash and ground ash. The raw low quality fly ash particle is coarse and irregular and its sieve residue of 45 μm reaches as high as 49.8%. The particle microstructure was significantly improved to be smaller and uniform after grinding. Table 2 presents the physical properties of the admixtures. Water demand for the raw fly ash is so high, because there is less free water in raw ash; on the other hand, there are more coarse pore carbon particle in raw ash.

Table 2. Physical properties of mineral admixtures.

Test item	Residue on sieve of 45 μm/%	Water requirement/%	Specific area/$m^2 \cdot kg^{-1}$	Density/$kg \cdot m^{-3}$
Cement	—	—	437	3010
Limestone powder	16.0	98.0	780	2670
Raw fly ash	49.8	99.6	278	1880
Ground fly ash	19.7	100.4	794	2340

Figure 1. Particle morphology of raw ash and ground ash.

Particle sizes of cement, limestone, raw ash, and ground fly ash were analyzed by OMEC lazer particle size analyzer LS-C (III) and Figure 2 shows the cumulative diameter percentage and the differential curve. It can be seen that the particle size distributions of limestone powder and cement were very similar with the former being slightly smaller.

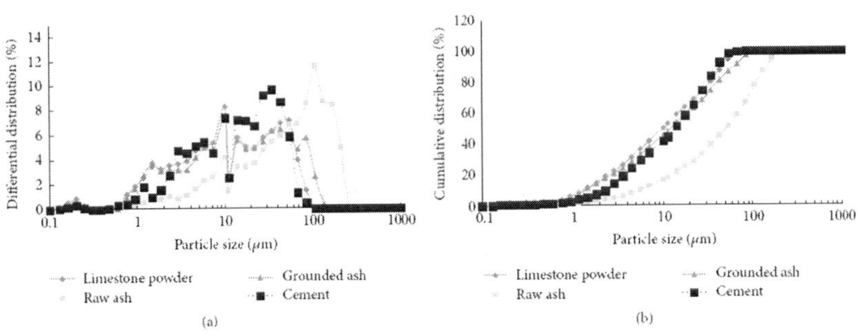

Figure 2. (a) Differential and (b) cumulative particle size distribution of the cementitious materials.

2.2. Testing Methods

The quantitative analysis of the influence of limestone powder and low quality fly ash on the strength and microstructure of cement-based materials was based on the comparison of pure cement paste with the content of 20%, 40%, and 60%. Water-binder ratio is 0.5. Table 3 lists the paste mix proportions.

Table 3. Mix proportion of composite cementitious materials/mass, %.

Sample	Cement	Limestone powder	Raw fly ash	Ground fly ash
C	100	—	—	—
L2	80	20	—	—
L4	60	40	—	—
L6	40	60	—	—
R2	80	—	20	—
R4	60	—	40	—
R6	40	—	60	—
G2	80	—	—	20
G4	60	—	—	40
G6	40	—	—	60

Paste specimens with 40 mm × 40 mm × 40 mm dimensions and a 0.4-water-binder ratio (W/C) were molded and cured with higher than 90% relative humidity and °C temperature until the stipulated age of 3, 7, 28, and 90 days. Determining the compressive strength of mortar specimens of different ages was conducted by WAY-2000 (i.e., a battery solution type compressive testing machine). Clean, bean-size samples were taken from the center of broken specimens and packed into ampere bottles filled up with absolute ethanol to terminate the hydration process before the XRD, TG-DTA, and SEM microcosmic analysis. Sample pieces were held in a dry environment at 60°C for 2 or 3 hours before SEM tests. The other samples were ground in an agate mortar and dried at 60°C for 2 hours to reduce carbonation before the XRD test and then dried in vacuum condition before the TG-DTA test.

The XRD analysis was made by X-ray diffraction, using a copper target and a continuous scan machine produced by RIGAKU, a Japanese company. The TG-DTA in an N2-atmosphere up to a temperature of 1200°C adopted the diamond TG/DTA analysis produced by Perkin Elmer Instruments Plant, and the morphology of the products was investigated using a scanning electron microscopy (JSM-5610LV, Japan).

3. RESULTS AND DISCUSSION

3.1. Strength

Figures 3 and 4 present the compressive strength of mortar at different water-binder ratios of 0.3 and 0.5. The figures showed that the strength of mortar decreased with increasing incorporation of admixture at the same water-binder ratio. The mortar containing low content of limestone powder showed higher compressive strength than that containing ground fly ash at the early stage of hydration. However, with the onset of hydration, the strength of the mortar containing ground fly ash improved rapidly and exceeded the one with limestone at the later age, which may be linked to the higher pozzolanic activity of fly ash at the later period.

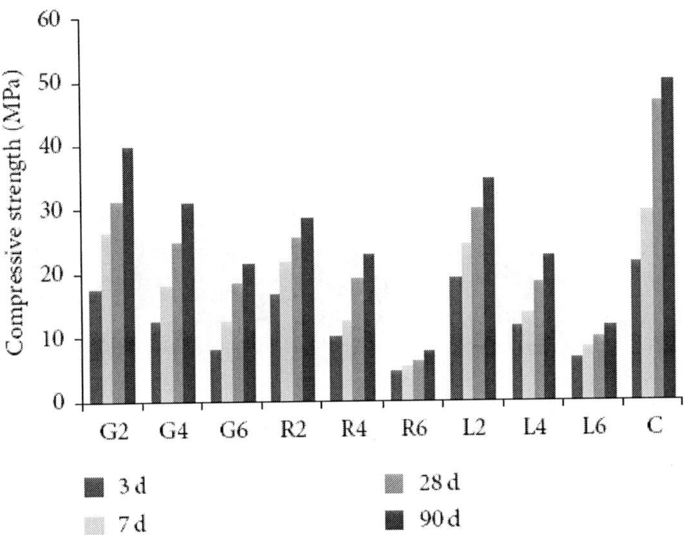

Figure 3. Compressive strength of mortars with water-binder ratio of 0.5.

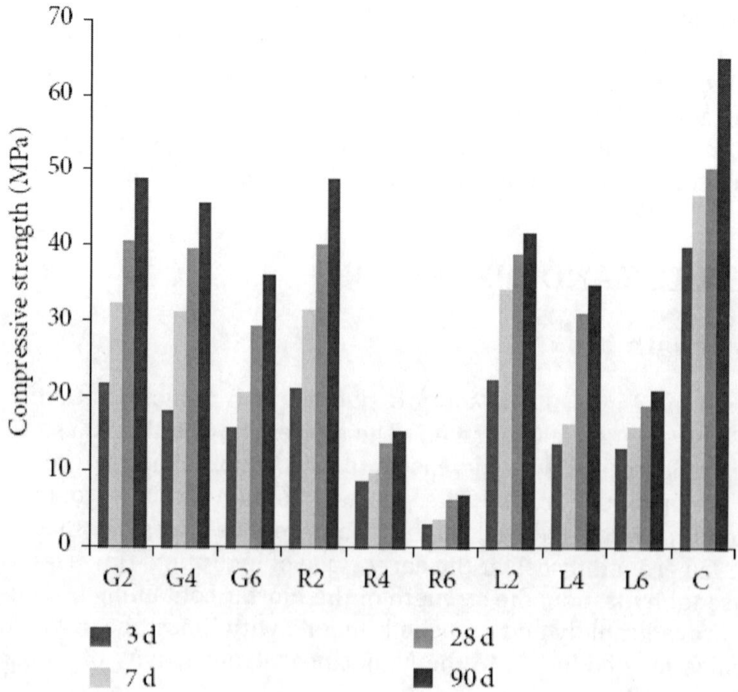

Figure 4. Compressive strength of mortars with water-binder ratio of 0.3.

The water-binder ratio had little influence on the strength developing law of mortar containing limestone powder or ground fly ash. Therefore, reducing the water-binder ratio could significantly improve the strength of mortar specimen containing high content of ground fly ash or low content (20%) raw fly ash to higher strength than those containing the same content of limestone powder.

3.2. Hydration Heat Evolution Process

Figure 5 shows the hydration heat produced and the rate of heat evolution for the four selected mixtures within the first 72 h. The hydration heat curves indicated that the influence of incorporating different admixtures in the composite cementitious materials varied the hydration heat. The addition of fly ash exhibited little effect on the five stages of cement hydration but reduced the hydration rate and the total heat compared to those of the pastes C. The third exothermic peak, caused by the conversion of AFt to AFm as a result of the consumption of gypsum, was much higher than the second one in paste G2 whereas the opposite response was observed in paste R2. This suggested that fine-grinding leads to a quicker hydration reaction and both raw ash and ground ash as mineral admixtures reduce the heat of hydration which is beneficial to mass concrete construction. The rate of heat evolution of pastes L2 exceeded that of G2 after the induction period but became lower during the deceleration period. It is proposed that the transformation from ettringite to a more steady monocarbon aluminate contributed to the third exothermic peak.

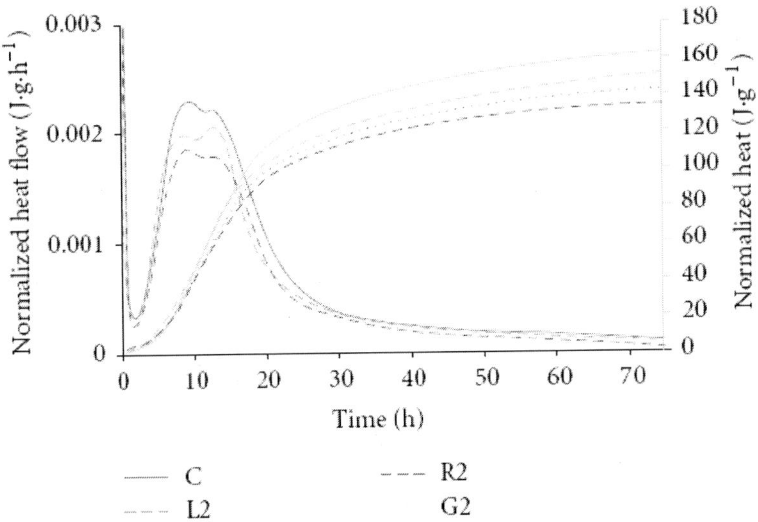

Figure 5. Hydration heat and rate of the cementitious materials.

For further analysis, Figure 6 presents a plot of the induction period against the acceleration period in the hydration process. The addition of fly ash prolonged the induction period but shortened the acceleration period compared with pastes C, which demonstrated that fly ash had little effect on the appearance of the second exothermic peak. There are two reasons for the extension of the induction period. On one hand, the increase of effective water reduced the concentration of Ca^{2+} in the pore solution and extended the induction period by increasing the time required for Ca^{2+} saturation. On the other hand, ground fly ash sphere particles may adsorb Ca^{2+} in solution, which inhibited the increase of Ca^{2+} concentration through adsorption in the first few hours. This also delayed the crystallization nucleation of CH and C-S-H and accordingly cements hydration. In the accelerated phase, more water came into contact with the new exposed surface and accelerates cement hydration because of the increasing effective water. In summary, the nucleation effect of the ground fly ash may also accelerate cement hydration.

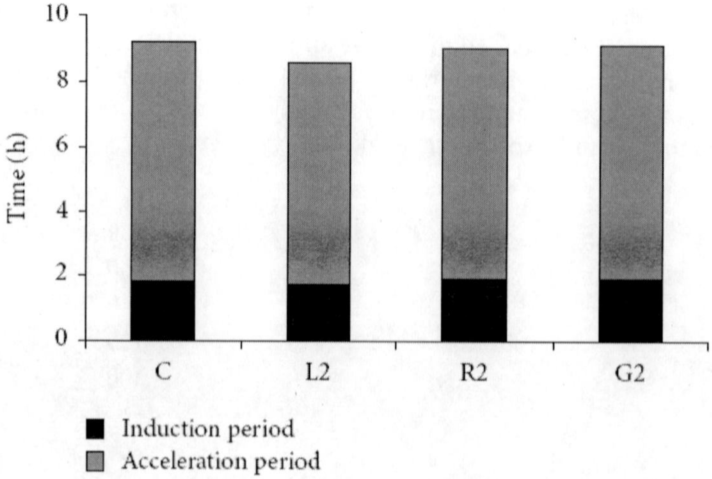

Figure 6. Induction period and acceleration period of the cementitious materials.

The difference between pure cement and cement-based materials is that the cement-based material containing limestone powder had three special points. These special points included a shortened induction period, an accelerated period, and a significantly higher third exothermic peak at around 13 hours. It accounted for the assertion that limestone powder does not take part in hydration due to its inertia but acts as a filler and contributes to the nucleation of C_3S and C_2S [22].

3.3. Microstructure

3.3.1. Hydration Products

Figures 7 and 8 show the XRD patterns of the hydration products of the selected pastes at 7 and 90 d, respectively. The main peaks in pastes C are those of calcium hydroxide (CH) and ettringite. The peaks of mullite (M) and quartz (Q) in the specimen containing fly ash are clearly evident, indicating that fly ash did not influence the types of cement hydration products. This was because the main chemical composition and component were the same. In addition to the noticeable diffraction peaks of calcium carbonate, there were also obvious single carbon calcium aluminate diffraction peaks in pastes L2. This was due to the calcium carbonate which reacted with C_3A to produce single carbon hydrated calcium aluminate as the main ingredient of limestone powder.

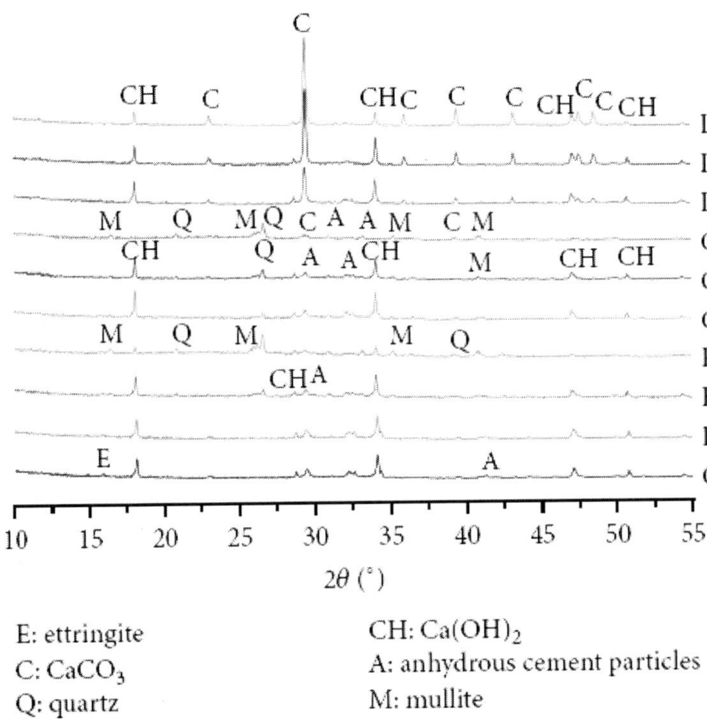

Figure 7. XRD patterns of specimens at 7 d.

Figures 9 and 10 also showed that the CH content in pastes R2 was lower at 90 d, indicating that the raw ash consumed some CH at 7 d. As for pastes G2, grinding may promote the activity, which made much higher consumption of CH at the secondary hydration, causing CH content at 90 d to be far less than that at 7 d. In addition, due to the consumption of cement clinker with hydration process, the characteristic peak of clinker was also reduced, which indicated exactly that fly ash exhibited a filling effect in promoting the hydration of cement to form CH. However, the ground one worked better. The CH content in pastes R2 was slightly higher at 90 d, suggesting that the raw ash was less active than the ground ash for the coarse particle, contributed to a relatively loose mortar structure. The strength of pastes R2 was lower.

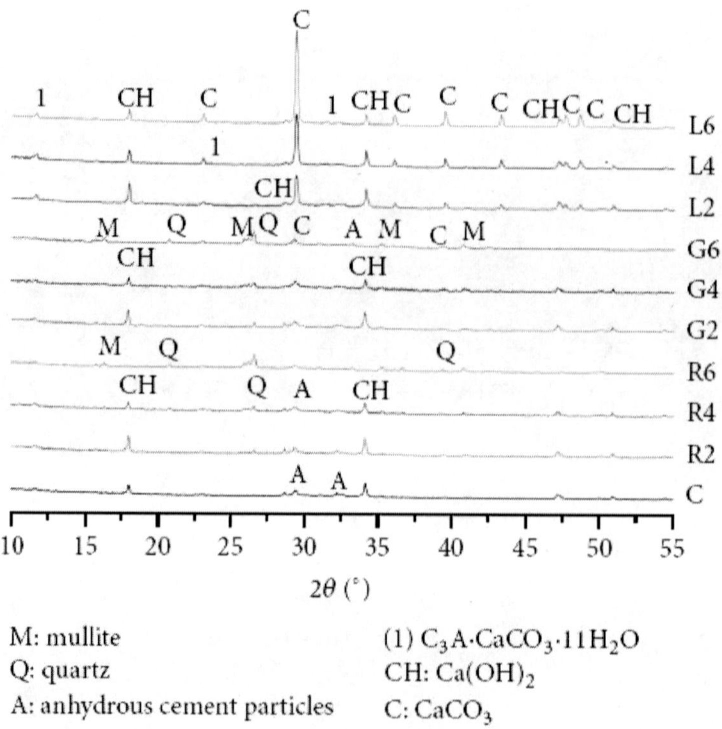

Figure 8. XRD patterns of specimens at 90 d.

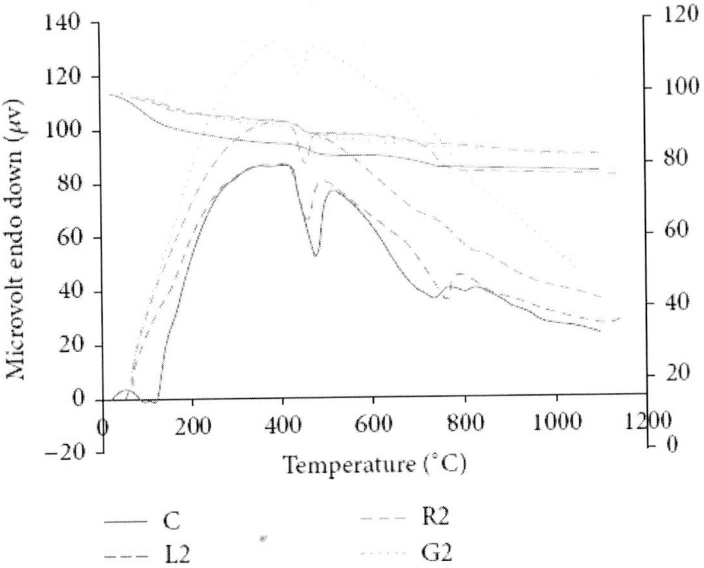

Figure 9.TG-DTA curves of C, L2, R2, and G2 at 7 d.

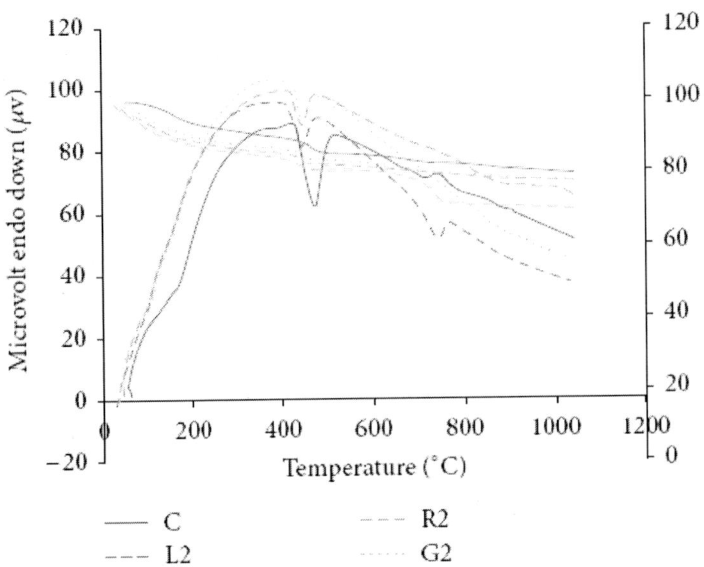

Figure 10. TG-DTA curves of C, L2, R2, and G2 at 90 d.

When the content of fly ash was up to 40%, the calcium hydroxide content was lower than that of the cement paste at the age of 90 d and was significantly lower than that of the sample at the age of 7 d. As a result, the early filling effect of fly ash promoted the cement hydration; however, the late pozzolanic reaction was gradual and consumed part of the calcium hydroxide. Comparing each group with admixture content of 20%, the amount of calcium hydroxide decreased at the age of 90 d for two main reasons. The first reason was the relative decrease in the cement content and the amount of calcium hydroxide produced. The other reason was that the pozzolanic activity of ash was more evident with increasing ash. Moreover, from the slurry map of ash mixed with raw fly ash, the content of calcium hydroxide changed similarly compared with that of ground fly ash indicating that the raw fly ash had pozzolanic activity.

When the content of fly ash was up to 60% (high volume), the diffraction peak of calcium hydroxide was significantly lower than that of the pure cement hydration at the age of 7 d and 90 d, respectively. The content of CH was less or even disappeared at the age of 90 d compared with that at the age of 7 d because the CH content produced by the cement hydration failed to meet the needs of the pozzolanic reaction of ash.

The significant differences between the pastes fully demonstrated that limestone powder promoted cement hydration with growth in age. Raw ash exhibited a filling effect in promoting cement hydration for its coarse particle and low activity. Grinding improved the activity of fly ash and was also conductive to its pozzolanic activity.

The content of nonevaporable water mainly depends on the quantity of hydration products, which relates to the hydration degree. Therefore, the content of nonevaporating water can be used to characterize the degree of cement hydration. The content of nonevaporating water is calculated as follows:

$$H = w_{105} - w_{1050} - \gamma,$$

$$\gamma = p_f r_f + p_c r_c, \tag{1}$$

where is the mass fraction measured by TG at corresponding temperature, is the loss of ignition, is the mass fraction of admixture, is the mass fraction of cement and, and are the corresponding losses of ignition. Table 4 summarizes the results obtained from Figures 9 and 10.

Table 4: Nonevaporating water content and hydration degree/mass, %.

Sample	Nonevaporating water	
	7 d	90 d
C	13.88	16.62
L2	11.5	12.02
R2	12.34	13.35
G2	12.91	14.89
L4	8.51	10.26
R4	9.33	11.53
G4	11.01	12.31
L6	2.89	3.42
R6	3.35	4.83
G6	4.16	5.19

Fly ash carries out the secondary pozzolanic reaction and participates in the hydration and excitation of calcium hydroxide as hydration product of cement clinker, contributing to strength. So the amount of CH consumed can indirectly reflect the hydration degree and the TG method can accurately determine the content of CH. The calcium hydroxide content is calculated according to the following equation and results given in Table 5:

$$\mathrm{CH} = \left(w_{\text{initial}} - w_{\text{finish}}\right) \cdot \frac{74}{18}. \tag{2}$$

Table 5: CH content of the hydration products/mass, %.

Sample	Seven days	90 days
C	13.46	16.39
L2	13.36	14.81
R2	12.53	11.35
G2	12.18	11.04
L4	12.01	10.88
R4	11.98	10.65
G4	9.59	9.37
L6	4.13	3.99
R6	2.92	0.85
G6	1.23	0.00

Table 5 lists the CH content calculated according to TG curves.

At 7 d, the nonevaporating water in the group with admixture incorporated was lower than that in group C, which showed that admixtures promoted cement hydration to varying degrees. The nonevaporable water content of groups G2 and R2 did not appear to be much different from each other and was just slightly higher than that of group L2. This was due to the low content of cement and the higher activity of G2 than that of R2, which accounted for the large number of nonevaporating hydrogels. In contrast, the limestone powder only promoted cement hydration. At 90 d, the content of nonevaporating water of each group differed a lot and increased with age. The content of nonevaporating water of group L2 remained stable and low, which means that the later products of L2 contained less bounded water. Meanwhile, the nonevaporating water of the pastes containing fly increased significantly, indicating that hydration products between 7 and 90 d have a rapid growth and form gel containing more nonevaporating water.

Table 5 indicated that CH content of group C and L2 pastes increased as hydration degree increased gradually, whereas that of the pastes containing fly ash decreased. From our previous analysis, it can been seen that the nonevaporating water of pastes C and the pastes containing fly ash increased greatly while that of pastes L2 tended to be stable. Since limestone powder primarily played the filling role, the content of CH consumed by fly ash in the two reactions was larger than that produced by the cement hydration. Thus, the nonevaporating water of C-S-H gel was much higher. The CH content of group R2 was significantly less than that of group L2 and obviously decreased further in group G2, which is consistent with the XRD results (i.e., remind us very briefly of what XRD says). The phenomenon indicated that grinding stimulated pozzolanic activity of fly ash effectively. With the content of limestone powder increase and cement content decrease, hydration product CH is from cement so that the CH content decreases for L4 and L6 samples. The CH content decreases with hydration age (because of the decrease of cement) and limestone content. As for R4, raw fly ash content is 40% which would decrease the hydration product CH. Otherwise, as the secondary hydration between fly ash and CH, the pozzolanic activity of raw fly ash is not as good as ground fly ash; the values of sample G4 were so low.

3.3.2. Morphology of Hydration Products

As shown in Figure 11, at 28 d, a large number of fibrous C-S-H gels grew as clusters of radiation and overlapped in group C. The microstructure of the pastes containing raw fly ash was loose and the particle surface was substantially smooth. The clubbed C-S-H gel and irregular gel as growing hydration products attached to the particles can be observed. The observation means that fly ash mainly played the role of filling at an early stage of hydration. A large number of C-S-H gels were produced in group G2 and made the structure denser. The main hydration products of group L2 were CH, ettringite crystal, and C-S-H gel of type II netlike and type III granular; at that time, limestone powder particles were already wrapped by a lot of dense gel.

C-S-H gel and thin layered CH staggered (see Figure 12); hence the structure was denser in group C at 90 d. Moreover, a large amount of layered CH could be observed in group R2; fly ash particles were wrapped in lots of hydration products, which showed that fly ash had been involved in the hydration limitedly. This is basically the same as group G2, but the C-S-H gel which wrapped the fly ash particles at the later stage was denser. It could also be observed that the surface of the limestone powder particle had been eroded and the structure of the reaction product was relatively denser, indicating that the limestone powder had taken part in the hydration process.

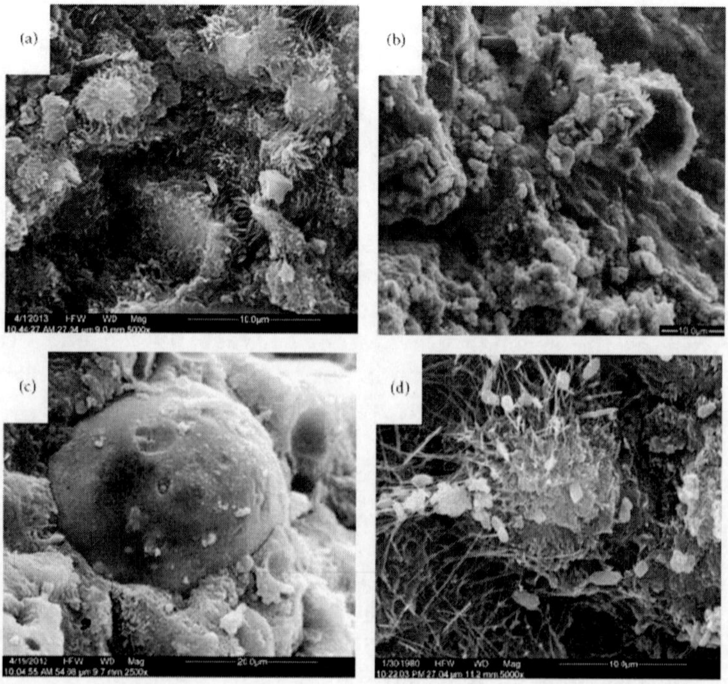

Figure 11: Microstructure of hydration products of (a) C, (b) L2, (c) R2, and (d) G2 at 28 d.

Due to the thick ash, the surface of the glass beads was substantially smooth at the age of 28 d (see Figure 13). Also it was observed that increased hydration products presented in the surroundings and are attached to the surface of the structure that were relatively loose, illustrating that the fly ash primarily acted as microaggregate filler. The fly ash particles appeared as etching and a dense layer of hydration products covered their surfaces at the age of 90 d. Again, a large amount of tufted C-S-H became very compact and dense and grew to intermingle with a laminar sheet of CH forming denser microstructure dense. Although the pozzolanic reaction of fly ash had been clearly evident, the consumption of CH was not much and the flaky CH crystals were still abound. Even in the slurry sample with up to 60% of fly ash, a large number of laminated growth of CH crystals existed next to the fly ash particles with a considerable degree of reaction. This observation indicated that the consumption of CH in fly ash reaction was small, where more of the CH crystals were still in the slurry cementitious material with high volume of fly ash to maintain the internal alkaline slurry at a higher level.

Figure 12: Microstructure of hydration products of (a) C, (b) L2, (c) R2, and (d) G2 at 90 d.

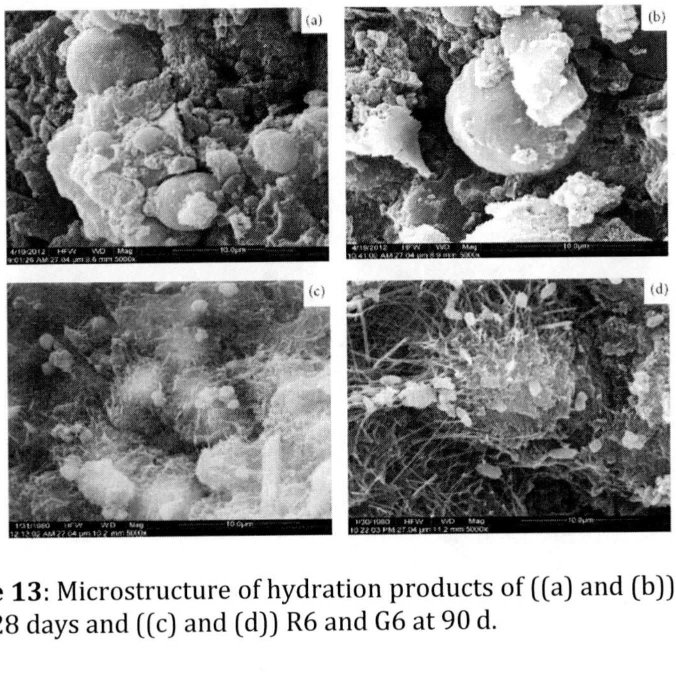

Figure 13: Microstructure of hydration products of ((a) and (b)) R6 and G6 at 28 days and ((c) and (d)) R6 and G6 at 90 d.

4. CONCLUSIONS

The results from this study support the following.

The strength of mortar decreased with increasing limestone powder content. The water-binder ratio had little influence on the development rule of the strength of mortar containing limestone powder. Limestone powder mainly played roles in promoting cement hydration and it hardly participated in hydration reaction. However, at the later period (90 d), limestone powder showed weak chemical activity. It may react with C_3A to generate monocarboaluminates.

Compared with limestone powder, low quality fly ash may improve the mechanical properties of cement-based materials and ameliorate the hydration and microstructure. In addition, low quality fly ash showed better mechanical property at a low water-binder ratio. The early hydration heat of the binder containing low quality fly ash was lower than that of the binder containing limestone powder.

Grinding did not only affect the physical properties but also improve the hydration of fly ash in cement-based materials. At the early age, the activity of low quality fly ash was low; but, at the later stage, it exhibits improvements, especially after grinding. Furthermore, low quality fly ash delayed the induction period, shortened the accelerator period, and hence had little effect on the second exothermic peak.

COMPETING INTERESTS

The authors declare that they have no competing interests.

ACKNOWLEDGMENTS

This material is based upon work supported by the Central Nonprofit Scientific Research Fund for Institutes (no. CKSF2016003/CL); National Natural Science Foundation of China (51208391 and 51139001), China Postdoctoral Science Foundation (no. 2015M582213) and National Basic Research Program of China (973 Program, 2013CB035901).

REFERENCES

1. P. K. Mehta, "Reducing the environmental impact of concrete," Concrete International, vol. 23, no. 1, pp. 61–67, 2001.

2. E. Worrell, L. Price, N. Martin, C. Hendriks, and L. O. Meida, "Carbon dioxide emissions from the global cement industry," Annual Review of Energy and the Environment, vol. 26, pp. 303–329, 2001.

3. S. H. Liu and P. Y. Yan, "Effect of limestone powder on microstructure of concrete," Journal Wuhan University of Technology, Materials Science Edition, vol. 25, no. 2, pp. 328–331, 2010.

4. S. H. Liu, P. Y. Yan, and J. W. Feng, "Effect of limestone powder and fly ash on magnesium sulfate resistance of mortar," Journal of Wuhan University of Technology (Material Science Edition), vol. 25, no. 4, pp. 700–703, 2010.

5. D. Zorić, D. Lazar, O. Rudić, M. Radeka, J. Ranogajec, and H. Hiršenberger, "Thermal conductivity of lightweight aggregate based on coal fly ash," Journal of Thermal Analysis and Calorimetry, vol. 110, no. 1, pp. 489–495, 2012.

6. T. Zhang, Q. Yu, J. Wei, P. Zhang, and P. Chen, "Improvement of surface cementitious properties of coarse fly ash by dehydration and rehydration processes," Journal of Thermal Analysis and Calorimetry, vol. 109, no. 1, pp. 265–271, 2012.

7. A.-M. Poppe and G. De Schutter, "Cement hydration in the presence of high filler contents," Cement and Concrete Research, vol. 35, no. 12, pp. 2290–2299, 2005.

8. V. Rahhal and R. Talero, "Early hydration of portland cement with crystalline mineral additions," Cement and Concrete Research, vol. 35, no. 7, pp. 1285–1291, 2005.

9. K. A. Paine, L. Zheng, and R. K. Dhir, "Experimental study and modelling of heat evolution of blended cements," Advances in Cement Research, vol. 17, no. 3, pp. 121–132, 2005.

10. P. Lawrence, M. Cyr, and E. Ringot, "Mineral admixtures in mortars: effect of inert materials on short-term hydration," Cement and Concrete Research, vol. 33, no. 12, pp. 1939–1947, 2003.

11. P. Y. Yan and Q. H. Zhang, "Influence of active or inert mineral admixtures on the hydration properties of composite cementitious materials," Journal of Railway Science and Engineering, vol. 4, no. 1, pp. 1–7, 2007.

12. P. Yan and Q. Zhang, "Microstructural characteristics of complex binder paste containing active or inert mineral admixtures," Journal of the Chinese Ceramic Society, vol. 34, no. 12, pp. 1491–1496, 2006.

13. W. Wongkeo, P. Thongsanitgarn, P. Chindaprasirt, and A. Chaipanich, "Thermogravimetry of ternary cement blends: effect of different curing methods," Journal of Thermal Analysis and Calorimetry, vol. 113, no. 3, pp. 1079–1090, 2013.

14. J. Baron and C. Douvre, "Technical and economical aspects of the use of limestone filler additions in cement," World Cement, vol. 18, no. 3, pp. 100–104, 1987.

15. M. Heikal, H. El-Didamony, and M. S. Morsy, "Limestone-filled pozzolanic cement," Cement and Concrete Research, vol. 30, no. 11, pp. 1827–1834, 2000.

16. B. Lothenbach, G. Le Saout, E. Gallucci, and K. Scrivener, "Influence of limestone on the hydration of Portland cements," Cement and Concrete Research, vol. 38, no. 6, pp. 848–860, 2008.

17. X. T. Huang and Z. J. Han, "Research and application of influence of man-made sand powder on the properties of concrete," Hydroelectric Engineering, vol. 49, no. 1, pp. 32–35, 1995.

18. X.-G. Li, "Use of man-made sand with high content stone powder in concrete," Journal of Building Materials, vol. 7, no. 1, pp. 66–71, 2004.

19. M. K. Zhou, S. M. Peng, and J. Xu, "Effect of stone powder on stone chippings concrete," Journal of Wuhan University of Technology (Material Science Edition), vol. 11, no. 2, pp. 328–331, 1996.

20. S. H. Liu and K. H. Fang, "Review on comprehensive utilization of fly ash," Fujian Building Materials, vol. 37, no. 2, pp. 8–9, 2008.

21. Y. Xi, "Present status and prospects of fly ash in comprehensive utilization," China Resources Comprehensive Utilization, vol. 27, no. 3, pp. 15–18, 2009.

22. G. Ye, X. Liu, G. De Schutter, A.-M. Poppe, and L. Taerwe, "Influence of limestone powder used as filler in SCC on hydration and microstructure of cement pastes," Cement and Concrete Composites, vol. 29, no. 2, pp. 94–102, 2007.

10

SLIDING WEAR PERFORMANCE EVALUATION OF RED MUD (RM), RM + FLY ASH (FA) AND RM + FA + AL COATINGS ON MILD STEEL

Harekrushna Sutar[1,2,3*], Debashis Roy[2], Subash Chandra Mishra[3], Rabi Ranjan Murmu[1]

[1]Chemical Engineering Department, Indira Gandhi Institute of Technology, Sarang, India

[2]Chemical Engineering Department, Jadavpur University, Kolkata, India

[3]Metallurgical and Materials Engineering Department, National Institute of Technology, Rourkela, India

ABSTRACT

The research paper focuses on evaluating the dry sliding wear behaviour of plasma sprayed coatings like red mud (RM), RM + 5% Fly Ash (FA) and RM + 5% FA + 5% Al on mild steel cylindrical shape substrate. Spraying was done at 10 kW operating power level. A conventional pin on disc wear tester is adopted for wear analysis. The wear test was performed at track diameter of 50 mm and sliding speed of 60 rpm (0.157 m/s) with normal load of 10 N. The duration of sliding varies from 54 minutes for pure red mud coating to a maximum value of 102 minutes for RM + 5% FA + 5% Al composite coating. A significant decrease in wear rate (in terms of mass loss) is observed by reinforcement of fly ash and aluminium to red mud, which might be due to improved interfacial properties.

Keywords: Red Mud, Fly Ash, Aluminium, Plasma Coating, Sliding Wear, Frictional Force

1. INTRODUCTION

Coating technology is an emerging research in engineering application. Wear resistive materials are fabricated by means of coating for wide tribological applications. Plasma spraying technology is widely adopted for coating thereby enhancing the surface properties. Surfaces with coating can protect from abrasive, adhesive and corrosive wear mediums. Commonly conventional coating materials are nickel, iron, cobalt and molybdenum based alloys, and carbides of ceramic and tungsten [1] [2] .

Several factors may influence the tribological behaviour of a coated surface. Such as geometry of contact, material characteristics, basic mechanical properties, operating parameters controlling the coating deposition and finally the microstructure [3] . Ceramic coatings are also proved to be favourable towards better tribological application. Tribological performance at high temperature of ceramic coatings is available [4] , showing high wear resistance [5] . But ceramic coatings exhibit failure mechanisms like plastic deformation, brittle fracture and polish effect [6] , which in turn demands reinforcement of additives which could reduce the friction and wear of plasma sprayed ceramic coatings. The importance of mechanical and tribological properties has convinced many researchers to study the friction and wear behaviour and to improve the wear resistance of composite coatings [7] . Composite coatings can be considered to be one of the competitive materials for tribological applications because of their low friction values against steel counterparts, good damping properties, and self-lubricating abilities [8] .

The resources of red mud and its comprehensive utilizations are reviewed [9] . Red mud as an industrial waste material may be considered to be the material of choice for coating applications for replacing some conventional expensive coating materials. Few results on the basis of sliding wear behaviour of red mud were being reported [10] . Morphology and solid particle erosion wear behaviour of red mud and fly ash composite coatings were being available in the literature

[11] . Dry sliding wear behaviour of rice husk and red mud filled epoxy composites is described [12] . Friction and dry sliding wear behaviour of red mud filled banana fibre reinforced unsaturated polyester composites is reported [13] . Tribological studies on red mud reinforced aluminium metal matrix composites are disclosed [14] . The characteristics of plasma sprayed pure red mud coating surface are reported [15] . Wear behaviours of plasma sprayed red mud coatings made at 9 kW operating power level, reinforced with varying weight % of fly ash, aluminium, carbon have been analysed at different operating conditions [16] . The objective of this research work is to study the mechanical and dry sliding wear behaviour of red mud coatings reinforced with additives like fly ash and fly ash + aluminium separately and comparing the results, inciting red mud (with additives like fly ash and aluminium) as suitable coating material may be implemented for replacing the conventional coatings.

2. MATERIALS AND METHODS

2.1. Raw Materials, Sample Preparation and Coating

The primary raw material for the study is red mud (RM). The red mud powder was collected from National Aluminium Company (NALCO), Damonjodi, Odisha, India. The additive like fly ash was obtained from Rourkela steel plant. Aluminium powder was procured from Rourkela market. After obtaining the precursor, the same was dried to make it moisture free. Powders chosen for plasma spraying are tabulated in Table 1.

Each raw material was sieved to obtain a particle size range of 80 - 100 μm and thoroughly blended to obtain uniformity in composition. The substrate selected for plasma coating is of mild steel. Commercially available mild steel rod was chopped to pieces of cylindrical dimension having $l = 30$ mm and $\Phi = 10$ mm. Spraying was done on both the cross section of the specimen. The average roughness of the specimen cross section is 7.2 μm. Specimens were pre cleaned in ultrasonic cleaning unit before plasma spraying. The raw materials were thermally sprayed using atmospheric plasma spraying technique. The parameters governing the spraying process are reported in Table 2.

2.2. Friction and Wear Test Apparatus

Dry sliding wear tests were carried out using a pin on disc friction and wear monitor test ring (DUCOM) according to ASTM G 99 standard. The cylindrical substrates with coating on both the cross sections were tested against a disc made of hardened ground steel (EN-32, hardness 72 HRC, surface roughness Ra = 0.07 μm). The schematic diagram of the pin on disc wear tester is shown inFigure 1.

The cylindrical test specimen was held stationary and hardened disc rotates anticlockwise against it. A normal force was applied on the test specimen by means of pulley, lever and spring mechanism. All wear experiments were carried out at normal atmospheric temperature and pressure. A normal load of 10 N was applied on substrate keeping track diameter at 50 mm and sliding speed of 60 rpm (At a velocity of 0.157 m/s). Wear rates are expressed in terms of mass loss in mg. Frictional force (F) was calculated directly from the apparatus in Kg.

Table 1. Powders used for coating.

Sl. No	Coating powder	Composition by weight %
1	Red mud (RM)	100
2	RM + Fly ash (FA)	95 + 5
3	RM + FA + Al	90 + 5 + 5

Table 2. Variables during plasma sprayed coating.

Operating variables	Set point
Plasma arc current (ampere)	200
Arc voltage (Volt)	50
Torch input power (kW)	10
Plasma gas (Argon) flow rate (Litre/Min)	20
Powder feed rate (gm/min)	10
Torch to base distance (mm)	120
Arc length range (mm)	3, 6, 9, 13
Spray angle (In degree)	90

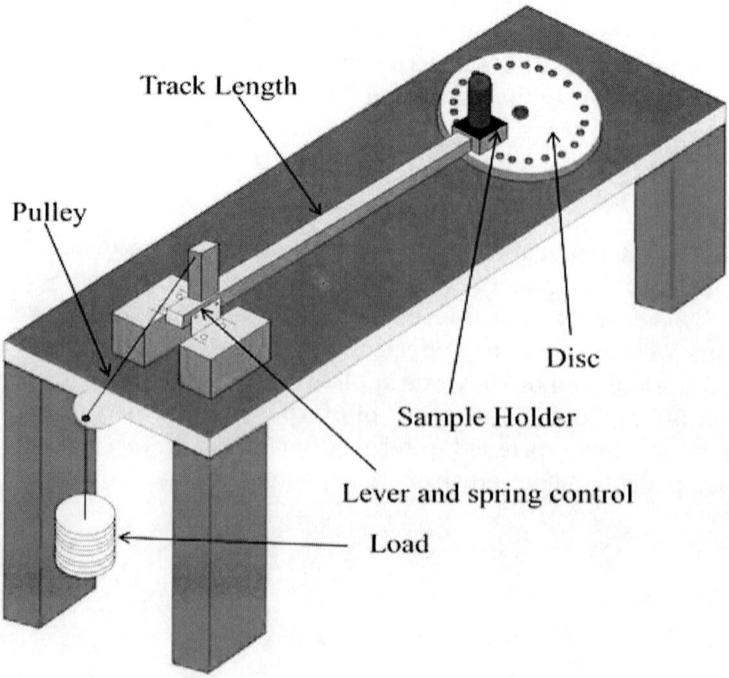

Figure 1. Model of pin on disc wear test rig.

3. RESULTS AND DISCUSSION

3.1. SEM, XRD and EDS Report

The Characterization of red mud powder was done using a SEM (Scanning electron microscope, JEOL; JSM- 6480 LV). The captured image and EDS (Energy dispersive spectroscopy) reports are shown in Figure 2.

The elemental analysis report is composed in Table 3 indicating weight (normalised) and atomic % (unnormalised) of elements comprising red mud powder. The major constituents prevailing in red mud are Fe and Ti. Minor constituents like Na, O, Mg, Si, K and Cr are also visible.

Further Characterization of red mud powder was carried by means of X-RD. The X-Ray diffractogram was captured using a Philips X-Ray diffractometer. The report is shown in Figure 3. The major phases found in red mud powder were Hematite (Fe_2O_3), Silicon dioxide (SiO_2), Titanium dioxide (TiO_2). The coatings are additionally recognized by means of EDS report. To perceive the significance; the EDS (Figure 4) and interlayer cross sectional sight (Figure 5) are reported for RM + 5% FA coating.

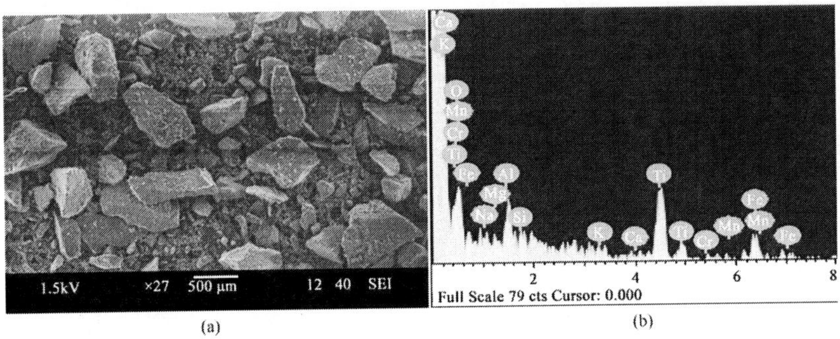

Figure 2. (a) SEM; (b) EDS of red mud powder.

Table 3. Elemental analysis report of red mud powder.

Element	Weight %	Atomic %
O K	0.70	1.90
Na K	1.77	3.35
Mg K	2.41	4.32
Al K	7.76	12.54
Si K	3.88	6.02
K K	0.71	0.79
Ca K	1.19	1.29
Ti K	46.63	42.44
Cr K	0.68	0.57
Mn K	1.76	1.40
Fe K	32.53	25.39
Totals	100	

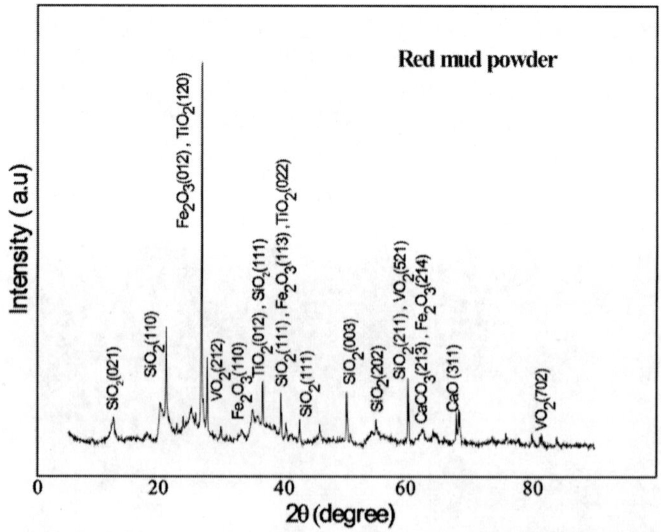

Figure 3. XRD of red mud powder.

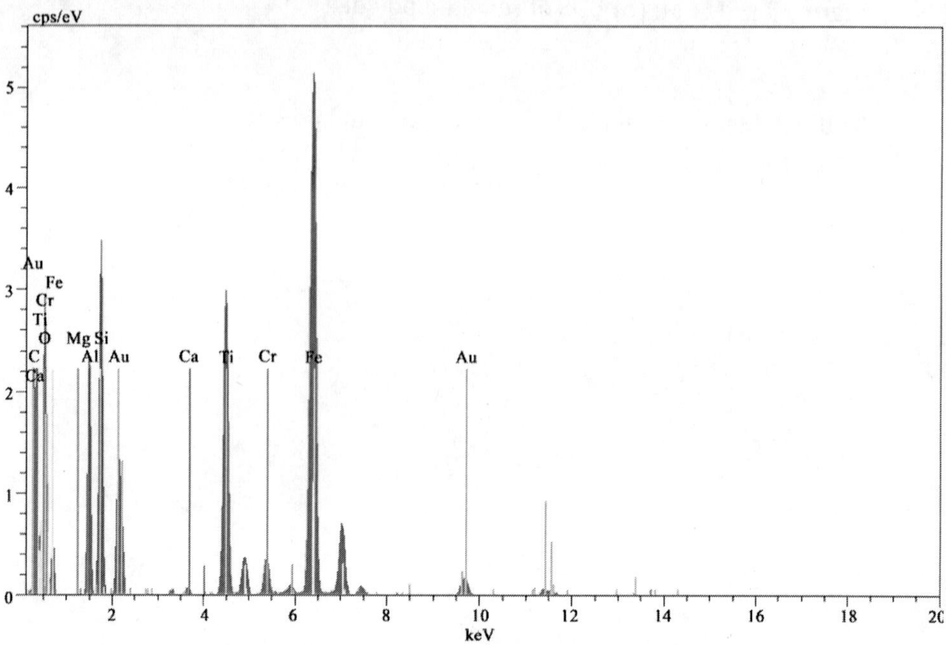

Figure 4. EDS outline for RM + 5% FA coating.

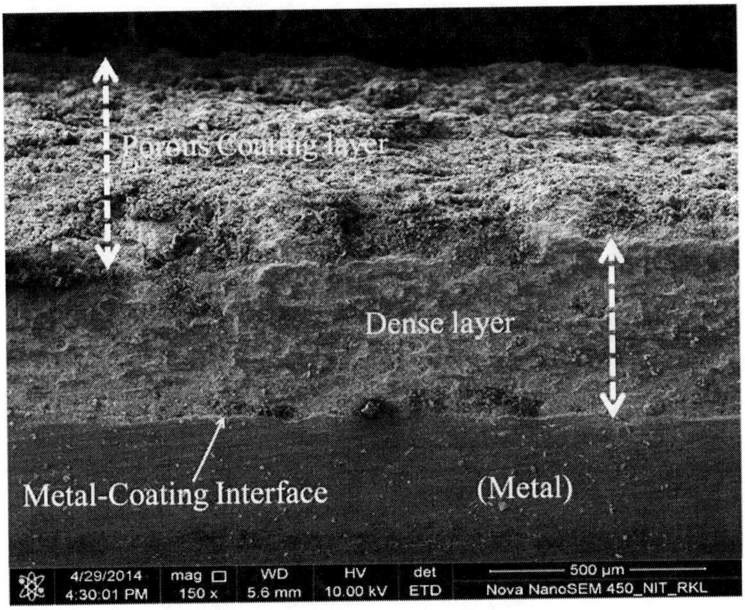

Figure 5. Snap shot of cross section for RM + 5% FA coating.

The morphology of each coating type was studied using FESEM (Field emission scanning electron microscope, NOVA Nano SEM-450). The captured images are shown in Figure 6. All the coatings were made at 10 kW operating power level. Figure 6(a) represents the morphology template for pure red mud coating. Very large size globular particles with non uniformity of interface were seen. Few semi molten small particles were also visible. Large scale cavity areas were found. As the composition changes to RM + 5% FA, very marginal changes in morphology pattern (Figure 6(b)) were observed. Similar cavities were found. But the numbers of semi molten particles were reduced. Smaller size globular particles were also visible. Finally as 5% Al was reinforced to RM + 5% FA composite, the coating surface format (Figure 6(c)) significantly affected. Uniformity of the surface was improved. Formation of cavity was fallen. But still very small size globular particles were seen.

Figure 6. Morphology of coatings (a) RM, (b) RM + 5% FA and (c) RM + 5% FA + 5% Al.

3.2. Wear Rate and Frictional Force Analysis

At first the coating surfaces were cleaned with acetone using tissue paper for removing the embedded particles if any. Then substrates were analysed for wear loss fixing in the pin holder. The operating parameters were secured at its values as mentioned earlier. Figure 7 shows the trends of wear rate for all coating during the course of examination. At the beginning the red mud coatings were examined. The wear analysis was performed till the coating layer prevails on the metal interface. The total duration of wear for red mud was continued up to 54 minutes as late as the existence of coating film. In each 3 minutes time gap the wear loss (In terms of mass loss in mg) was measured. A dramatic increase in mass loss was observed at incipient up to 36 minutes. There after the wear rate stabilises and the coating sheet terminates at 54 minutes.

An inclusion of 5% FA to RM reduces the wear rate as visible in Figure 7. A similar trend of higher mass loss at beginning was seen. The mass loss declines at the final stage and a steady state was attained at 54 minutes. The survival of coating surface ends at 72 minutes. Coatings of RM + 5% FA + 5% Al were audited eventually; designating a minimum mass loss. As before, substantially a higher tendency to wear was informed. A steady mass loss was occurred at 84 minutes and the wear rate plot assumed a plateau. The coating surface holds out at 102 minutes. The results of wear rate may be due to the distinction of coating inter layer properties, signifying the inequality of interfacial bond strength for each coating type. In other words for RM + 5% FA + 5% Al coating, there might be proper particle to particle bonding; which in turn augmented the interfacial tension, molecular bond strength as well as density and recasting to high wear resistive. The data pertaining to the force of friction (F) is reported in Figure 8.

Wear morphology for stipulated coating samples were captured. Figure 9 shows the wear morphology for RM + 5% FA coating after 15 and 30 minutes of sliding. Worn surfaces for RM + 5% FA + 5% Al coating at the end of 18 and 90 minutes sliding are pictured in Figure 10. Wear debris particles were visible on the worn surfaces with few cracked sections, winding up; the wear takes up by abrasive and adhesive mechanism.

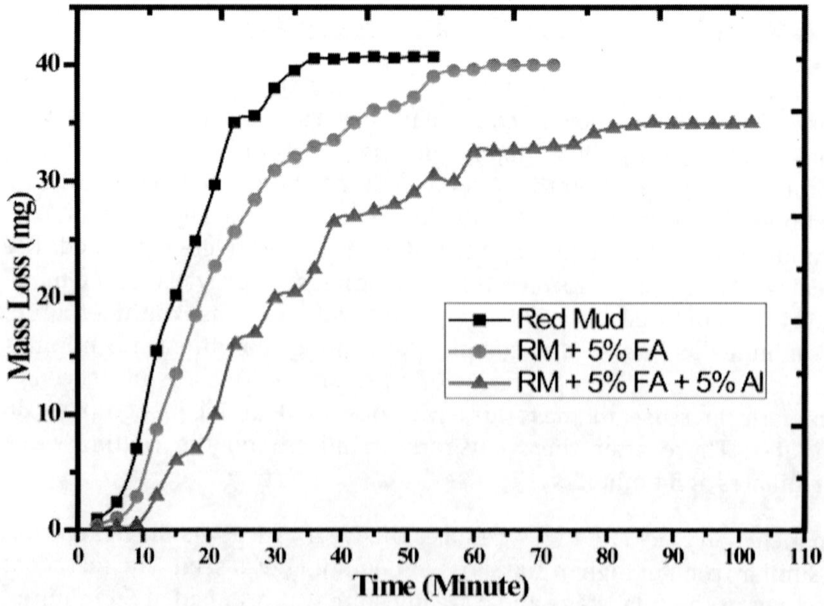

Figure 7. Wear rate plot for coatings.

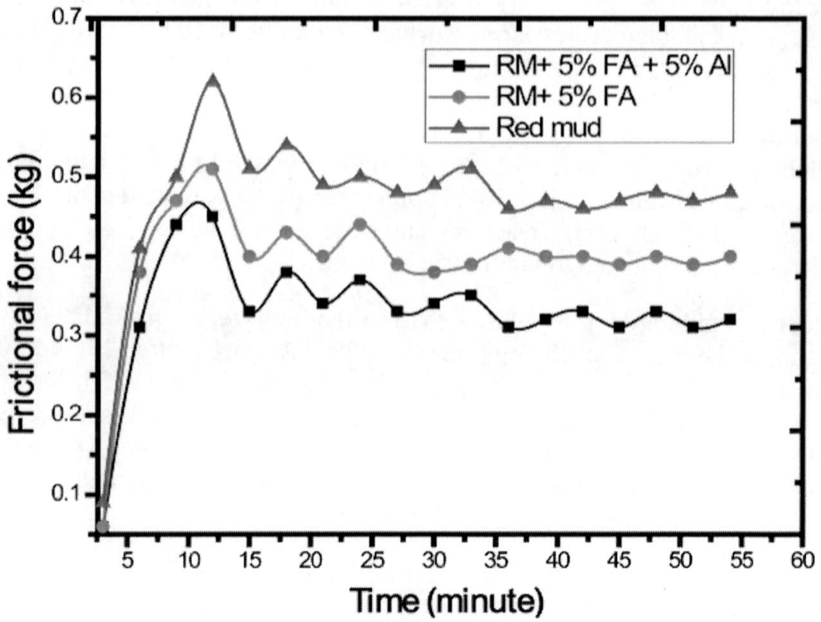

Figure 8. Variation of force of friction.

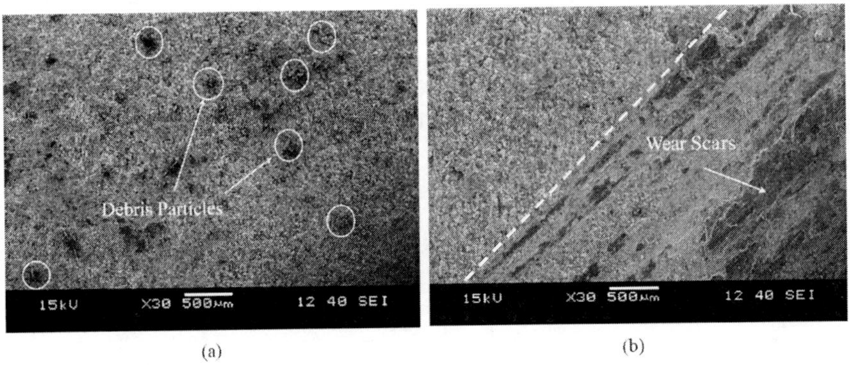

Figure 9. Wear morphology for RM + 5% FA coating after (a) 15 and (b) 30 minutes of sliding.

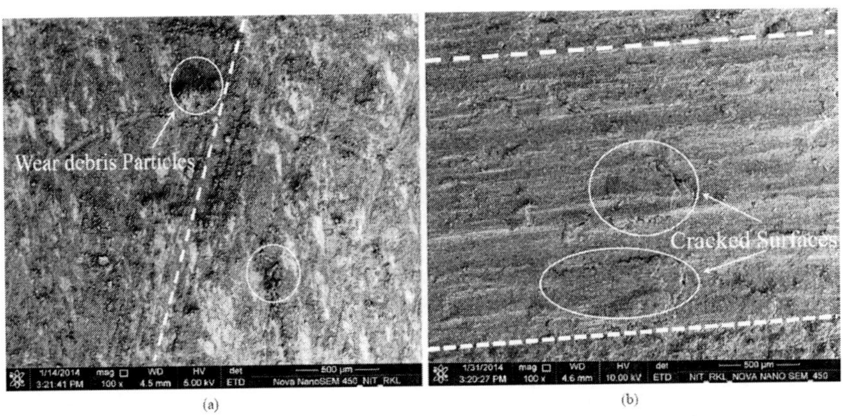

Figure 10. Worn surfaces for RM + 5% FA+ 5% Al coating after (a) 18 and (b) 90 minutes of sliding.

4. CONCLUSION

The above available research remarks a noteworthy accomplishment. The red mud, an industrial waste from alumina plant is proved to be conveniently coat able on metal like mild steel; and might be an alternative source for replacing conventional coating materials. Additives like fly ash and aluminium boost up the surface properties by modifying the physical and interfacial characteristics. The extended road map for the researchers could be to find out suitable tribological application areas. Studies on evaluating the resistance towards corrosive wear in different environmental conditions could be implemented. A subsequent inspection on the effect of heat treatment and high temperature application of these coatings may be emphasized.

REFERENCES

1. Kushner, B.A. and Novinski, E.R. (1992) Thermal Spray Coating. Friction, Lubrication and Wear Technology, 18, 829-833.

2. Crook, P. (1992) Friction and Wear of Hard Facing Alloys. Vol.18 of ASM Handbook, ASTM International.

3. Holmberg, K, Matthews, A. and Ronkainen, H. (1998) Coatings Tribology-Contact Mechanisms and Surface Design. Tribology International, 31, 107-120.

4. Wang, Y., Jin, Y. and Wen, S. (1988) The Analysis of the Friction and Wear Mechanisms of Plasma-Sprayed Ceramic Coatings at 450°C. Wear, 128, 265-276.

5. Wang, Y., Jin, Y. and Wen, S. (1998) The Analysis of the Chemicalstructure and Properties of Ceramic Surface Films in Friction Using SEM, AES and Micro-Region X-Ray Diffraction. Wear, 128, 277-290.

6. Vijande-Diaz, R., Belzunce, J., Fernandez, E., Rincon, A. and Perez, M.C. (1991) Wear and Microstructure in Fine Ceramic Coatings. Wear, 148, 221-233.

7. Kumar, S. and Chauhan, S.R. (2012) Mechanical and Dry Sliding Wear Behavior of Particulate Fillers CaCO3 and CaSO4 Filled Vinyl Ester Composites. International Journal of Composite Materials, 2, 101-114.

8. Lackner, J.M., Waldhauser, W., Major, L. and Kot, M. (2004) Tribology and Micromechanics of Chromium Nitride Based Multilayer Coatings on Soft and Hard Substrates. Coatings, 4, 121-138.

9. Sutar, H., Mishra, S.C., Sahoo, S.K., Chakraverty, A.P. and Maharana, H.S. (2014) Progress of Red Mud Utilization: An Overview. American Chemical Science Journal, 4, 255-279.

10. Sutar, H., Roy, D., Mishra, S.C., Chakraverty, A.P. and Maharana, H.S. (2015) Friction and Wear Behaviour of Plasma Sprayed Fly Ash Added Red Mud Coatings. Physical Science International Journal, 5, 61-73.

11. Sutar, H., Mishra, S.C., Sahoo, S.K., Satapathy, A. and Kumar, V. (2012) Morphology and Solid Particle Erosion Wear Behavior of Red Mud Composite Coatings. Natural Science, 4, 832-838.

12. Rout, A. and Satapathy, A. (2012) Analysis of Dry Sliding Wear Behaviour of Rice Husk Filled Epoxy Composites Using Design of Experiment and ANN. Procedia Engineering, 38, 1218-1232.

13. Prabu, V.A., Manikandan, V. and Uthayakumar, M. (2012) Friction and Dry Sliding Wear Behavior of Red Mud Filled Banana Fibre Reinforced Unsaturated Polyester Composites Using Taguchi Approach. Materials Physics and Mechanics, 15, 34-45.

14. Prasad, N., Sutar, H., Mishra, S.C., Sahoo, S.K. and Acharya, S.K. (2013) Dry Sliding Wear Behavior of Aluminium Matrix Composite Using Red Mud an Industrial Waste. International Research Journal of Pure and Applied Chemistry, 3, 59-74

15. Satapathy, A., Sutar, H., Mishra, S.C. and Sahoo, S.K. (2013) Characterization of Plasma Sprayed Pure Red Mud Coatings: An Analysis. American Chemical Science Journal, 3, 151-163.

16. Sutar, H., Roy, D. and Mishra, S.C. (2015) Effect of Fly Ash and Carbon Reinforcement on Dry Sliding Wear Behaviour of Red Mud. Indian Journal of Materials Science, 2015, Article ID: 296324, 7 p.

11

STRENGTH DEVELOPING IN CLAY-ANDESITE BRICK

Víctor Julio García[1,2], Alonso R. Zúñiga-Suárez[2], Carmen Omaira Márquez[1,3], Juan G. Pérez[2], Francisco Fernández-Martínez[2,4], Francisco Hernández-Olivares[2,4]

[1]Universidad Nacional de Chimborazo, Riobamba, Ecuador

[2]Universidad Técnica Particular de Loja, Loja, Ecuador

[3]Universidad de Los Andes, Mérida, Venezuela

[4]Universidad Politécnica de Madrid, Madrid, España

ABSTRACT

The objective of this work was to develop a more and better understanding of the strength developing in clay-andesite fired brick. The purpose was to improve the quality of the clay bricks that are handmade in the southern region of Ecuador to make them suitable for more widespread use in the local construction industry. To achieve our goal, we first physically, chemically, and mechanically characterized the "clayey" and "sandy" materials used in the fabrication of handmade bricks in the region. Second, the optimal mixture (OM): the optimal proportion between the amounts of "clayey" and "sandy" material was sought. Third, clay bricks were prepared using the OM, baked at 950°C, and characterized. In addition, bricks produced by regional artisans were characterized, and the results were compared with the results obtained for the bricks prepared using the OM. Our data reveal that the optimal mixture is 50% "clayey" material and 50% "sandy" material and that with this mixture, an average improvement in brick quality of 300% can be achieved; thus, the use of the OM makes it possible to expand the use of these bricks in the local construction industry and enables an environmentally friendly production process by reducing the intensive exploitation of regional clay deposits. This improvement is achieved by virtue of the anorthite enrichment that occurs in the solid solution, which results from the evolution of andesine. The anorthite contributes to the formation of a stronger matrix among the different grains of the material. Conditions are favorable for this enrichment process to occur when "sandy" materials with high contents of andesite, which is common in Ecuadorian soils, are used.

Keywords: Anorthite, Andesine, Andesite Component, Fired Bricks, Brazilian Test

1. INTRODUCTION

Solid clay bricks are one of the oldest and most extensively used construction materials in human civilization. Clay bricks are formed through a simple manufacturing process (mixing, kneading, drying, and baking) and are an affordable product that is created using non-renewable resources, such as clay, which exists in abundance in diverse regions of the planet. Clay bricks offer solidity, durability, and compactness, among other potentially beneficial physical, mechanical, and thermal properties. However, clay brick technology has evolved very little throughout history, and efforts to improve its technological performance have received little attention. However, due to their low production costs and wide range of properties, clay bricks offer remarkable potential for technological development, although there is little fundamental knowledge available regarding the relationship among the composition, microstructure, and mechanical performance of such bricks [1] . For this reason, there is considerable interest in using technology to develop a new type of clay brick with improved performance in response to the technological challenges posed by the modern construction industry. However, the production of clay bricks has a high environmental impact in terms of the consumption of non-renewable resources (materials), energy, and water as well as the emission of greenhouse gases and the generation of waste, among other aspects [2] - [4] .

The materials used in the clay brick industry are dispersed throughout the earth's surface and have their natural origins in complex geological processes that occurred during the formation of the earth's crust. The possibility of using these materials in the manufacture of clay bricks depends primarily on the geological and geochemical conditions in which the materials are formed, as well as on their mineralogical composition. The environmental impact due to the consumption of non-renewable resources is related to the availability of these clayey materials and to the exploitation of areas that must remain available for the conservation of biodiversity. Thus, the intensive exploitation of clay deposits poses a threat to environmental sustainability and also threatens the development of a sustainable construction industry. The mitigation of the environmental effects of clay brick production and the technological development of optimized clay bricks begins with the formulation of an OM, considering the compositions of both the "clayey" material and the available "sandy" material. The dominant component in an OM is quartz, but the most important ingredients are hydrated silicates or clay minerals, which endow the mixture with viscosity and plasticity once it is hydrated. Other ingredients include carbonates, iron oxides, and other minerals and their mixtures, which are used as fluidifiers or degreasers, among other purposes. The mineral compositions of these clayey minerals vary greatly. The amounts and compositions of these minerals directly influence the molding and drying of a brick before it is placed in the oven as well as its behavior during baking and the color and quality of the finished product.

However, the most suitable proportion of "clayey" and "sandy" materials depends, to a large extent, on the nature of the available materials. In addition, we must consider that the requirements for environmental preservation are continuously increasing [2] . Therefore, the optimization of the ratio of "clayey" material to "sandy" material represents an effective and important means of mitigating the environmental impact [5] . At the Andes mountain range, "sandy" material is found in surface deposits that usually have good amount of andesite. Actually, andesite most commonly denotes fine-grained and typically consist essentially of andesine (a plagioclase feldspar). Crystal of andesine lies in a finer groundmass, usually crystalline, but sometime glassy. Andesine consists of 70% to 50% albite and 30% to 50% anorthite. Andesine can decompose into anorthite and albite, as occurs in the regional model of thermal metamorphism [6] [7] . However, studies of the formation of anorthite and their effects on the increase in strength during sintering of clay based material are rare in the literature [8] .

The objective of this work was to develop a more and better understanding of the strength developing in clay-andesite fired brick.

The purpose was to improve the quality of the clay bricks that are handmade in the southern region of Ecuador to make them suitable for more widespread use in the local construction industry.

To achieve our goal, we first physically, chemically, and mechanically characterized the "clayey" and "sandy" materials used in the handmade production of bricks in the region. Second, the optimal proportion between the amounts of "clayey" and "sandy" materials was sought and determined. Third, clay bricks were prepared using the OM and sintered at 950°C, and these bricks were then characterized. In addition, bricks produced by seven major artisanal producers from the region (with production capacities greater than 10,000 units) were characterized, and the results were compared with the results obtained for the bricks prepared using the OM.

2. MATERIALS AND METHODS

2.1. Characterization of Raw Materials ("Clayey" and "Sandy" Materials)

a) Granulometric analysis was performed using number 20, 40, 60, 100 and 200 mesh sieves to determine the particle size distribution [9] .

b) The geotechnical properties of the prepared pastes were then determined. The plastic limit (PL) was determined using the 3-mm-thread method, and the liquid limit (LL) was determined using a Casagrande device [10] . The plasticity index (PI) was calculated based on the arithmetic difference between the LL and the PL. For brick preparation, the mixture was moistened to five percentage points above its LL.

c) The raw material samples were subjected to differential scanning calorimetric-thermo gravimetric analysis (DSC-TGA) using an SDT Q600 device (manufactured by the Thermal Analysis Company) equipped with a platinum-platinum/rhodium thermocouple. The analysis was performed between room temperature and 1000°C at a heating rate of 20°C/min and under an oxygen flow. To achieve better peak resolution, a small amount of each sample was used (approx. 46 mg) and rapid heating was applied.

d) Chemical analysis (oxide percentage (wt %)) of the samples was performed using X-ray fluorescence (XRF) spectroscopy (S1 TURBO^{SD} handheld XRF analyzer, Bruker). For this analysis, the samples were mechanically ground in a planetary mill and reduced to powder in an agate mortar.

e) The crystalline phases were analyzed via the powder method using an X-ray diffractometer (D8 Advance, Bruker) equipped with a copper anode ($\lambda = 1.5418 \times 10^{-10}$ m) and a linear detector (LYNXEYE compound silicon strip detector, Bruker). The diffractograms were collected in a range of 5 to 70 degrees using a step size of 0.02 degrees and a measurement time of 2 seconds per step. The diffractogram analysis was performed through a comparison with the Joint Committee on Power Diffraction Standards spectra using the DIFFRAC.SUITE EVA software from Bruker.

f) The porosity, density, amount of water absorbed, and dimensional variation of the samples were determined as follows.

The shrinkage of each sample after being dried (DS) in an oven was determined using Equation (1):

$$\mathrm{DS}\ \% = \frac{L_1 - L_2}{L_1} \times 100\ \% \quad (1)$$

In Equation (1), L_1 represents the length of the specimen after drying under ambient conditions, before being placed in the oven. The shrinkage of each sample after being baked and sintered in the oven (FS) was determined using Equation (2):

$$\mathrm{FS}\ \% = \frac{L_2 - L_3}{L_2} \times 100\ \% \quad (2)$$

In Equation (2), L_2 represents the length of the specimen after drying in the oven, before being placed in the baking oven. L_3 represents the length of the specimen after being baked or sintered in the oven. The total shrinkage (TS) was calculated as follows:

$$\mathrm{TS} = \mathrm{FS} + \mathrm{DS} \quad (3)$$

The density and porosity were determined using the gravimetric method based on Archimedes' principle. The specimens were saturated with water for 24 h and were then weighed while immersed in water

(W_1) and in air after the surface water had dried (W_2). Subsequently, the samples were oven-dried for 6 h at a temperature of 105°C to completely remove the water, after which the dry specimens were weighed (W_3).

The amount of water absorbed (WA) was determined using Equation (4), following the American Society for Testing and Materials (ASTM) standard [11] . This amount corresponds to the relative amount (by weight percent) of WA by the specimen while immersed under the prescribed conditions with respect to the weight of the dry specimen:

$$WA(\%) = \frac{W_2 - W_3}{W_3} \times 100\% \quad . (4)$$

The fraction of open pores (f_o), or the apparent porosity [12] , was calculated using Equation (5):

$$f_o = \frac{W_2 - W_3}{W_2 - W_1} \times 100\% = \left(1 - \frac{\rho_b}{\rho_S}\right) \times 100\% \quad . (5)$$

The density of the solid phase (ρ_s) and the apparent density (ρ_b) were calculated using the ratios given in Equation (6) below [13] :

$$\rho_s = \frac{W_3}{W_3 - W_1} \times \rho_w \; ; \; \rho_b = \frac{W_3}{W_2 - W_1} \times \rho_w \quad (6)$$

In Equation (6), ρ_w = 997.04 kg/m^3 is the density of water at 25°C. The percentage of mass lost during baking and sintering was estimated using the following Equation:

$$LOS = \frac{W_4 - W_3}{W_3} \times 100\ \% \quad . \quad (7)$$

In Equation (7), W_4 represents the weight of the dry specimen before being placed in the oven for baking and sintering.

g) The indirect tensile (IT) test (the Brazilian test) was performed on the baked and sintered disks at various temperatures. In this test, the disk is vertically compressed to indirectly generate a tensile stress in the horizontal direction. The maximum value of the applied load (P_{max}) is recorded when the failure of the disk occurs, and the indirect tensile strength (σ_{IT}) is calculated using Equation (8):

$$\sigma_{IT} = \frac{2P_{max}}{\pi D l} \quad (8)$$

In Equation (8), D represents the disk diameter, and l represents its thickness. The IT test offers several advantages with respect to other methods. The test configuration is simple, and cylindrical or disk-shape specimens are used. A large sample volume can therefore be tested using this method, and it is not sensitive to surface defects [14] .

h) The compressive strength of the bricks was determined following the ASTM standard [11] , and the tests were performed in quintuplicate. For these tests, a universal testing machine was used (Versa Tester, Soil Test Inc., USA), and the compressive strength was determined by dividing the maximum applied load by the area of the brick face subjected to the load.

i) For the drying of the bricks and disks, the samples were left in the molds for 12 h and were then dried for 72 h under ambient conditions. Subsequently, they were oven-dried at 105°C for 24 h. The samples were heated at a heating rate of 0.08°C/min from room temperature to 60°C (approx. 8 h) and were then maintained at that temperature for 3 h; subsequently, the temperature was further raised at the same heating rate to 105°C, at which they remained for 4 h. The oven was disconnected, and the samples were allowed to cool inside the oven until they reached room temperature.

j) In the baking and sintering of the bricks and disks, the disks were baked and sintered at 500°C, 700°C, and 950°C, whereas the bricks were baked only at 950°C. The baking and sintering were performed in an electric oven at a heating rate of 1.66°C/min until the final temperature was reached, and this temperature was then maintained for 3 h. Finally, the samples were allowed to cool freely via natural convection until they reached room temperature.

2.2. Experimental Design

The raw "clayey" and "sandy" materials used to produce the handmade bricks were obtained from the natural deposits at "La Vega (LVG)" and "Indiucho (IND)", respectively. Both deposits are located in the Catamayo Canton of Loja Province in Ecuador. The LVG deposit is located at a southern latitude of 4°00'39.7" and a western longitude of 79°22'43.51", and its geomorphological structure belongs to the Gonzanamá Formation. The IND deposit is located at a southern latitude of 4°01'47.4" and a western longitude of 79°21'54.5", between the Sacapalca and Loma Blanca Formations. The Gonzanamá Formation dates from the Paleocene and rests discordantly on the Sacapalca Formation; it predominantly consists of sedimentary volcanic rocks such as dark-brown clays, gray lutites, tuffaceous sandstones, and tuffs. It is believed that this formation has a thickness of approximately 1000 meters. The Sacapalca Formation dates from the Paleogene and exhibits a lithology of andesitic lavas and pyroclastic rocks. The lavas are resistant to erosion and are located in the highest parts of the Catamayo Basin, whereas the pyroclastic rocks correspond to andesitic tuffs, agglomerate tuffs, and subordinate agglomerates. The lavas are brown or gray andesitic porphyry, with a high content of mafic minerals, including enstatite, hornblende, and augite. The Loma Blanca Formation dates from the Oligocene-Miocene and rests discordantly on the Gonzanamá Formation; it essentially consists of an agglomerate basal portion followed by a sequence of stratified medium-grain white tuffs, whose layers have an inclination angle of approximately 124 degrees in the SE direction. The rocks of this formation are wrapped in a silty matrix containing clasts of mainly variegated tuffs with sizes ranging from millimeters to centimeters. The formation is superimposed with a sequence of agglomerate tuffs, tuffs, and pyroclastic flows, whose deposition environment is of the continental type [15] [16] .

The samples were sieved (sieve #18) to remove the inert components of the material, thereby obtaining grains of less than 1 mm. No specialized treatment, such as those reported by Kornilov and Luzin [17] , was used for the activation of the clays. We characterize the clayey (from LVG deposit) and sandy materials (from IND deposit) by doing granulometric analysis, XRF, differential thermo gravimetric analysis, X-ray diffraction and the geotechnical properties of the 90/10 mixture. The optimization of the mixture and its characterization were performed following the procedure outlined in Figure 1.

To determine the OM, a simplex-reticular method was used, and mixtures were prepared with the following ratios of clayey/sandy materials, expressed in terms of the total weight percentage of the mixture: 30 - 70, 40 - 60, 50 - 50, 60 - 40, and 70 - 30. The materials were carefully aerated and manually mixed. Water was added to each mixture until a uniform consistency was achieved. Then, the pastes were allowed to rest for 24 h in sealed bags (maturation process) to achieve the maximum possible plasticity. Subsequently, portions of each paste were placed in circular molds of 75 mm in diameter and 20 mm in thickness to obtain disk-shaped samples. Three replicates were prepared for each formulation, yielding a total of fifteen samples. To achieve soft extrusion, a weight of 500 g was placed on each sample for 60 min while it was in the mold. Subsequently, the fifteen samples were dried at 105°C and then baked and sintered at 950°C. The disks were subjected to the IT test at a loading speed of 0.010 mm/min. To determine the OM, we defined the optimization factor (f_{opt}) in a manner similar to the structural efficiency factor defined in [18] . Thus, we defined the optimization factor as the ratio of the maximum stress supported by the sample under indirect tension (σ_{IT}), expressed in MPa, to the sample weight (p), expressed in kg. Thus,

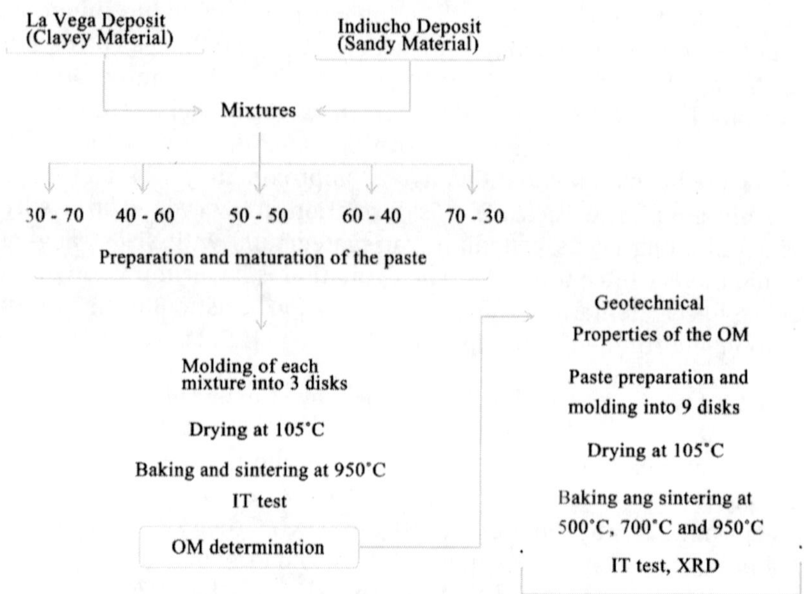

Figure 1. Methodology implemented to determine and characterize the OM. IT test represents the indirect tensile test.

$$f_{opt} \equiv \frac{\sigma_{IT}}{p}, \tag{9}$$

$$f_{opt} \equiv \frac{2}{\pi D l} \times \frac{P_{max}}{p}. \tag{10}$$

In Equation (10), D represent de diameter of the sample and l is its thickness. The OM was defined as the composition of the disk that exhibited the highest value of the optimization factor. Once the OM was identified, additional paste was prepared using that formulation. The geotechnical properties of the paste were determined, and it was used to prepare nine disk-shaped samples following the procedure described above. Once the samples were dried, they were divided into three sets of three. One set was baked and sintered up to a temperature of 500°C, the second was sintered at a temperature of 700°C, and the third was sintered at 950°C. The sintered disks were subjected to the IT test, and their chemical compositions were analyzed via X-ray diffraction.

Subsequently, sufficient paste to make five bricks was prepared by hydrating the OM paste to five percentage points above its LL. The paste was allowed to rest and then used to fill wooden molds of 270 × 140 × 80 mm. A weight of approx. 3 kg was placed on each mold to achieve soft molding or extrusion. The bricks were allowed to dry following the protocol described in section (2.1.i). Afterward, the bricks were baked and sintered following the method presented in section (2.1.j). Then, following the protocol described in section (2.1.f), conventional characterization (dimensional variation, water adsorption, density, and porosity) of the bricks prepared in the laboratory and the brick produced by the regional artisans was performed. Subsequently, the bricks were subjected to the direct compression test in accordance with the methodology defined in section (2.1.h).

The results obtained in the tests of the bricks prepared in the laboratory were compared with those obtained in the tests performed on the handmade bricks from the regional producers.

3. RESULTS AND DISCUSSION

3.1. Chemical Composition and X-Ray Diffraction

The chemical compositions of the LVG and IND samples are shown in Table 1. The analysis results are expressed as the percentages of oxides present, where the main chemical elements are Si, Al, Fe, and Ca. These oxide contents can be summarized as 56% to 61% SiO_2, 19% to 24% Al_2O_3, 3% to 8% CaO, 8% to 9% Fe_2O_3 + TiO_2, and 4% to 8% RO + R_2O. The diffractograms of the LVG and IND samples are presented in Figure 2, in which several of the most intense peaks are marked. A semi-quantitative analysis of the diffractograms shown in Figure 2 reveals that the LVG sample contained a mixture of illite (8%); muscovite-2 M (4%); kaolinite (18.3%); several plagioclases, such as sodium anorthite (3.6%), albite (5%), and andesine (7.56%); and quartz (41.76%). Meanwhile, in the IND diffractogram, the plagioclases sodium

Table 1. Chemical compositions of the LVG and IND samples determined via XRF.

	LVG (%)	IND (%)
SiO_2	56.52	60.02
Al_2O_3	23.08	19.28
CaO	3.51	7.64
Fe_2O_3	8.07	7.33
MgO	4.59	3.24
K_2O	2.42	0.82
MnO	0.17	0.20
TiO_2	0.75	0.77
Sb_2O_3	0.24	0.25

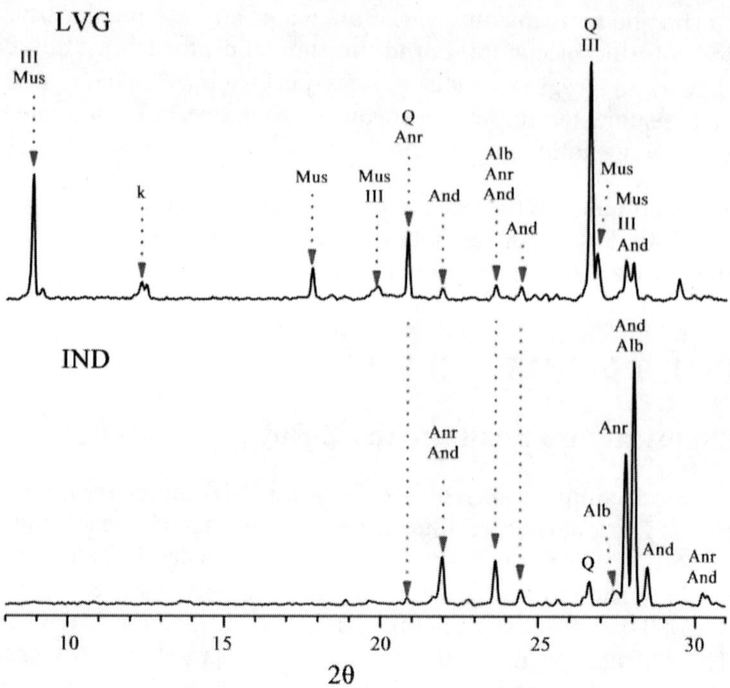

Figure 2. Diffractograms of the LVG and IND samples. Illite (Ill); muscovite-2M (Mus); kaolinite (k); anorthite (Anr), albite (Alb), and andesine (And); and quartz (Q).

anorthite (3.22%), albite (5.81%), and andesine (62.70%), as well as diopside (9.97%) and quartz (15.78%), predominate.

Adequate baking and/or sintering of a mixture used for brick fabrication requires knowledge of the thermal behaviors that the raw material can exhibit when subjected to a heating program. Such knowledge allows the changes in volume, phase, and crystallization, among others, that can occur during the heating process to be understood and anticipated. In addition, the most suitable baking temperature prior to vitrification can be determined. The thermal phenomena exhibited by a material as its temperature is increased, until it decomposes, depend to a large extent on the material's purity, its chemical composition, the nature of the bonds between its constituents (binding energy between atoms and/or molecules), and its surface properties. For example, clays with fine grains tend to decompose more readily than clays with coarse grains [19] .

3.2. Thermo Gravimetric Analysis

The temperatures at which the different thermal phenomena occur can be studied using thermal analysis methods such as differential thermal analysis, thermo gravimetric analysis (TGA), and differential scanning calorimetry (DSC). However, for samples that are a mixture of diverse materials, such as the materials used to prepare clay bricks, the interpretation of the data is complicated by the simultaneous occurrence of thermal phenomena of different origins. To facilitate the interpretation of the thermograms, we can use the first (D1) and second (D2) derivatives of the thermogram with respect to temperature to more clearly or explicitly visualize the occurrence of thermal events. The D2-DSC curve represents the changes in the heat flow rate of the sample during a range of diverse thermal processes, such as endo-and exothermic decomposition. The D2-DSC curve provides complementary information to facilitate better evaluation of the simultaneously occurring thermal processes recorded in the TGA and DSC thermograms. The thermograms of the LVG and IND samples are presented in Figure 3.

In the DSC curve shown in Figure 3(a), it can be observed that during the initial heating period, the sample released water in an endothermic reaction up to a temperature of approximately 120°C. In this process, the sample lost 4.58% of its mass, probably as a result of the removal of hygroscopic water from the clayey material

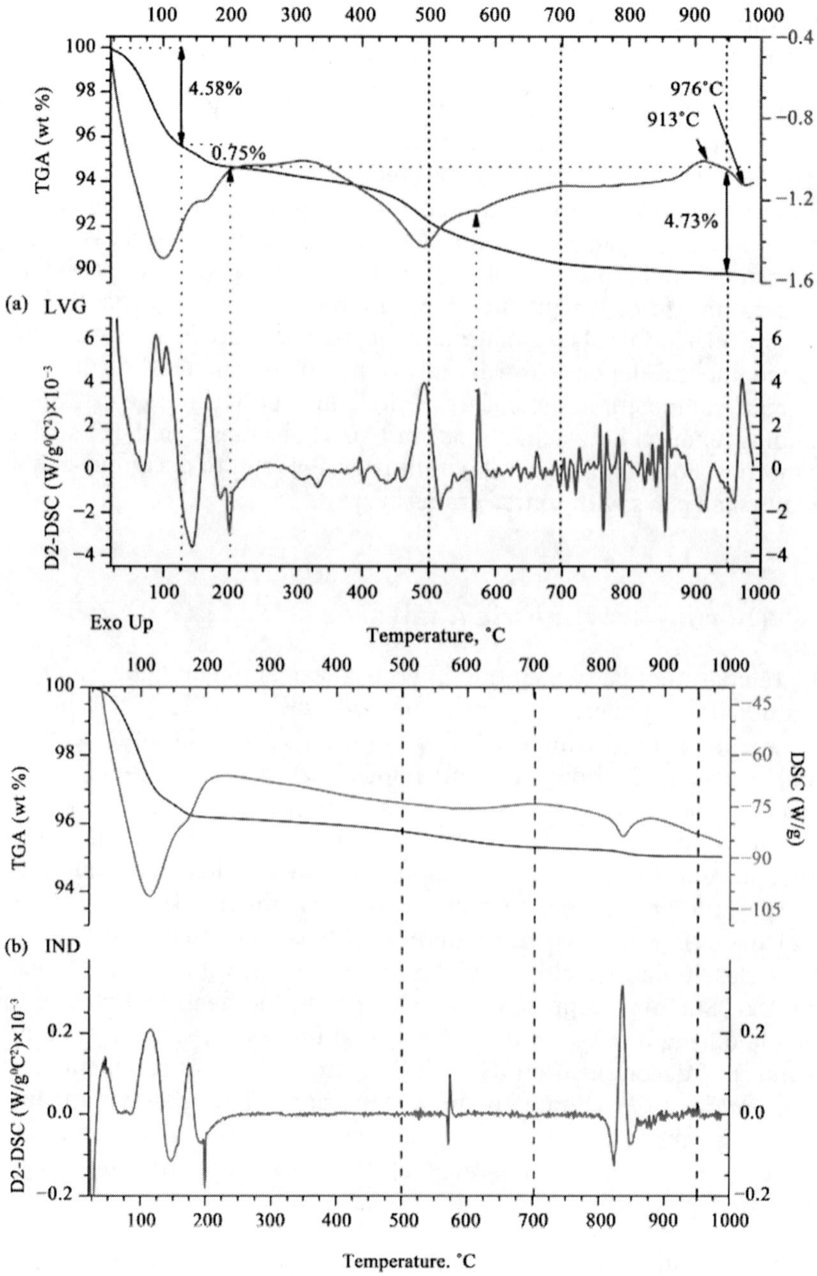

Figure 3. TGA, DSC, and D2-DSC thermograms of the LVG (a) and IND (b) samples. The dotted lines represent the final sintering temperatures.

(see the TGA curve in Figure 3(a)). The wet mass exhibited a dimensional change during drying at temperatures in the range of 0 to 150°C takes because of the loss of water from its pores [20] . Above 120°C and up to approximately 200°C, a turning point appears in the DSC thermogram ofFigure 3(a), probably related to the removal of hygroscopic water from the illite and other minerals, which caused the sample to lose approximately 0.75% of its weight. In the literature, it has been reported that the effect associated with kaolin in this temperature range is very small compared with that associated with illite [21] . At 200°C, a sharp endothermic peak is evident, probably due to the thermal desorption of weakly absorbed (physisorbed) water adhered to the particle surfaces through a van der Waals-type attraction. Such water is adsorbed at active sites associated with defects in the structure and is typically observed to be present in amounts directly proportional to the number of particles. In the range from 200°C to 500°C, the oxidation of organic matter typically occurs, and the loss of energy due to the loss of organic mass associated with the clay particles is relatively low [21] . In fact, the combustion of organic matter depends on the size of its constituent particles and on their distribution in the sample, which gives rise to small but wide peaks during heating. In the range of 300°C to 350°C, a maximum can be observed in the DSC curve in Figure 3(a), probably related to loss of water from the hydrated iron oxide [22] . In fact, a maximum can be observed in the D2-DSC curve in Figure 3(a) in the range of 304°C - 315°C, where the thermal energy flow remains constant over a small temperature range before it begins to decrease once again [21] . In the D2-DSC curve of Figure 3(a), between 380°C and 480°C, several events can be observed that are not visible in the DSC curve because they overlap with the organic matter decomposition; these events correspond to the dehydroxylation of the clay. Dehydroxylation is a thermally activated reaction and depends on many factors, of which time and temperature are the most important variables. This dehydroxylation is a process in which water molecules form and is removed from the structure of the clayey minerals, following the reaction $OH^- + OH^- \rightarrow H_2O + O^{2-}$ [23] . The dehydroxylation activation energies of the minerals decrease in the following order: illite, muscovite, montmorillonite, and kaolinite [24] ; this process is not clearly observed in other minerals, so it is not discernible in the curve in Figure 3(a).

Kaolinite is characterized by a prominent endothermic effect in the range of 450°C to 570°C, followed by the destruction of the mineral crystalline network and the formation of metakaolin accompanied by the release of water, following the reaction described in Equation (11) [25] - [27] . The formation of the water molecules generates voids in the crystalline network, and when sufficient voids have formed, the structure deforms, producing metakaolin:

$$Al_2Si_2O_5(OH)_4 \rightarrow Al_2Si_2O_7 + 2H_2O \tag{11}$$

The maximum of this endothermic event depends on the heating rate, occurring at 443°C when the heating rate is 1°C/min and at 527°C when the heating rate is 40°C/min [28] . The recrystallization of kaolinite occurs between 950°C and 1000°C, accompanied by a considerable release of energy, which should be observed in the thermogram as a sharp exothermic peak [20] [29] .

The loss of water from the illite results in an initial endothermic peak between 100°C and 150°C due to dehydration, followed by a second endothermic peak near 460°C - 470°C due to dehydroxylation and a third peak near 900°C that is associated with the final dehydroxylation stage [20] [30] . However, Gualtieri and Ferrari reported the existence of at least two endothermic peaks in the range from 600°C to 700°C in the dehydroxylation process of illite [24] [29] . In its final state, the onset of the dehydroxylation process occurs before the decomposition of the original silicate structure. The thermal decomposition of illite begins at 800°C, and as the temperature increases, hematite is formed at the expense of silicate [31] . At 910°C, the illite recrystallizes in an exothermic process [29] .

Muscovite-2M1 exhibits a significant phase change at approximately 800°C, which is attributed in the literature to a dehydroxylation process. The dehydroxylation of muscovite-2M1 occurs in the temperature range between 700°C and 1000°C and is accompanied by the nucleation and growth of the dehydroxylated phase at high temperatures [32] . Two thermal processes have been associated with the dehydroxylation of muscovite-2M1: one at 475°C and another at 750°C [33] . The process involves the condensation of two hydroxyl groups to form a water molecule via the following reaction:

$$KAl_2(Si_3Al)O_{10}(OH)_2 \rightarrow H_2O + KAl_2(Si_3Al)O_{11} \tag{12}$$

Anorthite (Anr), with the formula $CaAl_2Si_2O_8$, and albite (Alb), with the formula $NaAlSi_3O_8$, lie at the extremes of a continuous series of reactions in a solid solution of plagioclases, as described by Bowen. The intermediate members of the series are homogeneous mixtures of the products at the extremes and include albite, which consists of 100% to 90% Alb and 0% to 10% Anr; andesine, which consists of 70% to 50% Alb and 30% to 50% Anr; labradorite, which consists of 50% to 30% Alb and 50% to 70% Anr; and anorthite, which consists of 10% to 0% Alb and 90% to 100% Anr. This series of plagioclases forms a solid homogenous solution at high and low temperatures, and it arises through the complete substitution of sodium and silicon with calcium and aluminum.

Pure anorthite (Anr_{100}) exhibits a reversible structural phase transition (a primitive structure of space group $P\bar{1}$ to a body-centered structure of space group $I\bar{1}$) at approximately 237°C [34] .

Albite exists in two crystalline forms: a completely ordered form that is stable at low temperatures, called "low albite" or simply albite, and a disordered form that is stable at high temperatures, called "high albite". The transformation from albite to "high albite" occurs in the temperature range of 490°C to 670°C [35] .

Andesine can decompose into anorthite and albite, as occurs in the regional model of thermal metamorphism. Its homogeneous superstructure with periodic boundaries and non-periodic anti phase boundaries (APB) is locally distorted because of the energy accumulated along the closely spaced two-dimensional defects. Thus, the APB recombine in one region of the crystal to form anorthite (Anr_{95}), and in another region of the same crystal, the structure is transformed into albite ($Anr_{<5}$), without generating a superstructure [6] [7] .

At 573°C, alpha trigonal quartz transforms into beta hexagonal quartz. This transformation can be explicitly observed in the D2-DSC curve in Figure 3(a). The quartz content can be determined from the change in enthalpy, which, for pure quartz, is on the order of 7.5 J/s [33] . It can be expected that at 870°C, the beta quartz must irreversibly transform into beta tridymite, undergoing a significant change in volume (approximately 14.4%), but the kinetics of this reaction is very slow, and it never occurs below 1250°C. A gradual decomposition of quartz can also be expected in the range of 800°C - 1100°C, with a significant decrease in its content at 1100°C [21] . However, when impurities are present, the beta quartz can instead irreversibly transform into alpha cristobalite at 1050°C or 1250°C [36] .

All peaks related to the release of carbonate in the D2-DSC curve of Figure 3(a) must appear as narrow peaks in the temperature range of 640°C to 740°C [21] . Several processes occur in the range of 700°C to 900°C that are more visible in the D2-DSC curve in Figure 3(a) and are associated with the decomposition of the clays and of the calcite. Calcite can decompose over a wide temperature range, from 650°C (decarbonation of polycrystalline calcite) to approximately 850°C, at which the decomposition of monocrystalline calcite occurs. At 810°C, the decarbonation of carbonates with particle sizes similar to those of the clays occurs [29] .

The thermograms collected for the IND samples are shown in Figure 3(b), in which thermal events similar to those observed for the LVG sample are evident, particularly those in the temperature range between 120°C and 200°C. The endothermic peak at 573°C is related to the transformation of alpha quartz into beta quartz. However, in contrast to the thermograms of the LVG sample,Figure 3(b) exhibits a prominent endothermic peak at approximately 835°C, which can be attributed to sodium anorthite or K-feldspar [4] .

Finally, in Figure 3(a), it can be observed that at 976°C, vitrification of the clay occurred in the LVG sample, turning it into a fragile material. Therefore, we concluded that the most suitable baking temperature for clays of this type must lie between these last two temperatures, and thus, we selected the samples that were baked at 950°C for further examination.

3.3. Indirect Tensile Test and Structural Evolution

The IT test results for specimens with different proportions of IND and LVG materials baked at 950°C are summarized in Table 2. In the presence of a tensile stress, the disks prepared using the various mixtures behaved similarly to somewhat fragile materials such as concrete and rocks. The tensile strength of such materials is very low compared with the compressive strength and is frequently ignored. The tensile strength depends primarily on the strength of the mineral grains and on the matrix that holds them together. In addition, there is also some dependence on the chemical composition, the presence of inclusions, and the quality and dimensions of the pores. For this reason, the strength of such a material strongly depends on the strength of its weakest regions; the homogenous clayey material used as one of the raw materials, with its low content of impurities, provided a high tensile strength [37] .

Table 2. Results of tests performed to optimize the proportions of LVG and IND materials in the mixture using specimens sintered at 950°C.

LVG (%)	IND (%)	*P* (g)	*D* (cm)	*l* (cm)	P_{max} (KN)	σ_{IT} (MPa)	f_{opt} (MPa/kg)	*LOS* (%)	DS (%)	FS (%)
30	70	118.81	6.94	1.70	0.56	0.40	3.24	5.27	0.30	10.71
40	60	147.95	6.85	2.24	1.33	0.55	3.75	5.75	0.29	2.36
50	50	117.29	6.79	1.84	1.42	0.73	6.20	6.33	0.24	1.52
60	40	137.66	6.71	2.21	1.55	0.67	4.90	6.90	0.43	2.98
70	30	139.64	6.66	2.28	1.68	0.71	5.12	7.61	0.49	4.63

Figure 4(a) shows the values of the optimization factor defined in Equation (10) for each of the tested samples. The mixture with a ratio of 50 - 50 exhibited the highest optimization factor (6.20 MPa/kg) and also the lowest dimensional variations after drying (0.24%) and after baking and sintering (1.52%). Based on these results, we defined the OM as a mixture containing 50% of LVG material and 50% of IND material, or, in other words, with a proportion of 50 - 50. Figure 4(b) shows the diffractograms corresponding to three samples with different proportions of LVG and IND materials sintered at 950°C, in which it can be observed that the evolution kinetics of the three mixtures are complex and distinct. The diffractogram of the sample with a proportion of 30 - 70 suggests a predominance of andesine, albite, and quartz, whereas in the diffractogram of the sample with a proportion of 50 - 50, a predominance of labradorite, andesine, and quartz is observed. Finally, the diffractogram of the sample with a proportion of 70 - 30 indicates the presence of albite, labradorite, and quartz.

Figure 5(a) shows the variations in the indirect tensile strength and the loss on sintering for the OM as functions of the final sintering temperature. Figure 5(b) shows the diffractograms of samples prepared using the OM and sintered at 500°C, 700°C and 950°C. For a sintering temperature of 500°C, a predominance of andesine, albite, and anorthite is observed, whereas for 950°C, there is a predominance of labradorite, anorthite, and albite. Thus, anorthite plays a very important role in the consolidation of the solid brick matrix because the decomposition of andesine into anorthite and albite contributes to the emergence of labradorite and the enrichment of the matrix with anorthite. As was mentioned, in one region of the andesine crystal, the APB recombines to form anorthite, whereas in another region of the same crystal, the structure transforms into albite without a superstructure [6] [7]. Smith

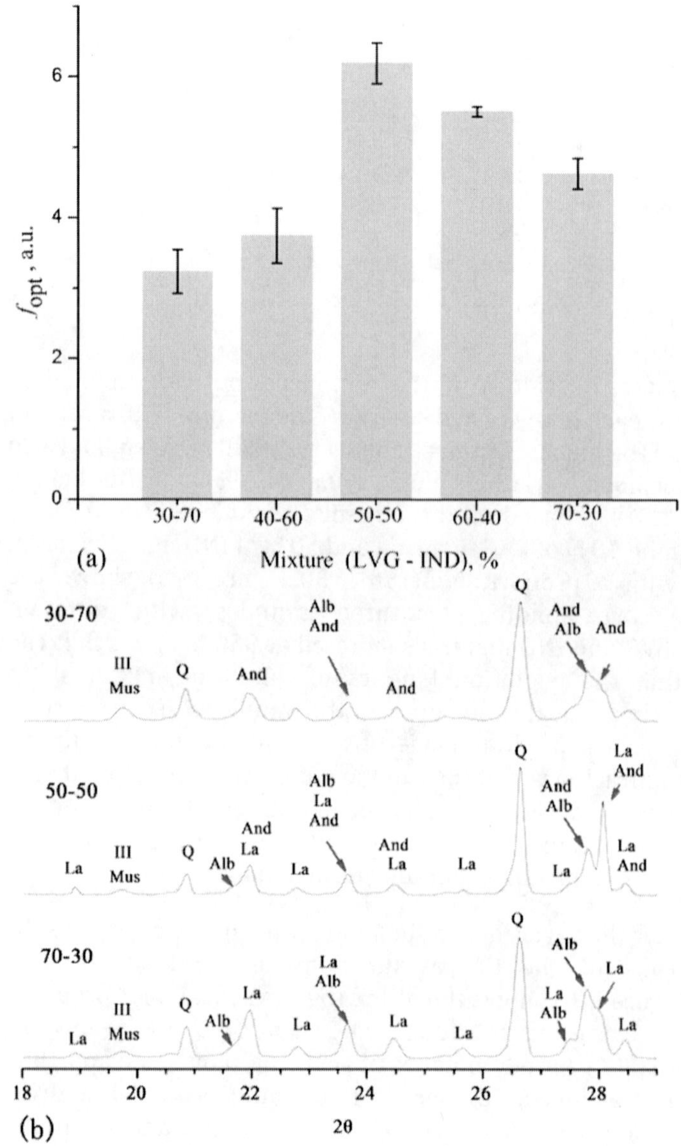

Figure 4. (a) Values of the optimization factor defined in Equation (10) for the various mixtures tested. (b) Diffractograms corresponding to samples with three different proportions of LVG and IND, sintered at 950°C. Illite (Ill); muscovite-2M (Mus); kaolinite (k); anorthite (Anr), albite (Alb), labradorite (La), and andesine (And); and quartz (Q).

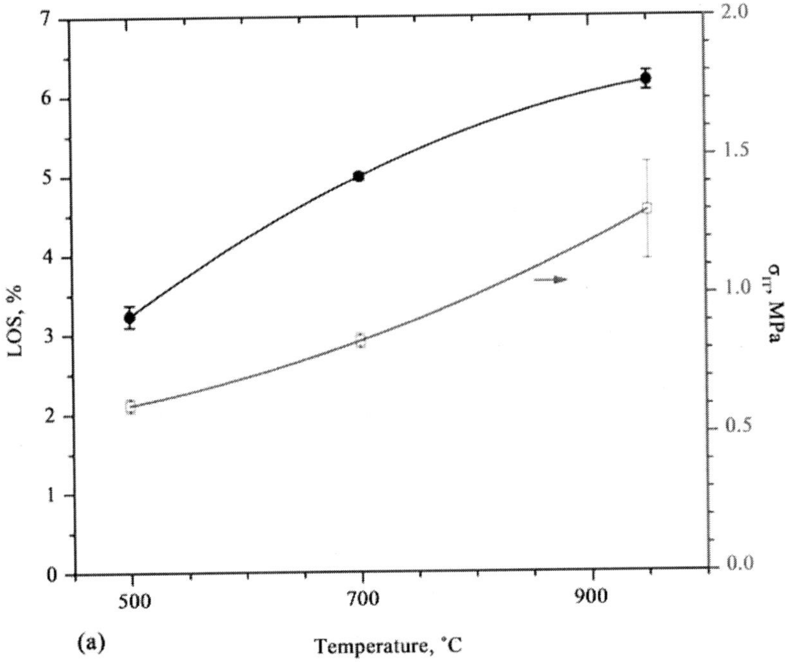

Figure 5. (a) Variation in the indirect tensile strength (σ_{IT}) of the OM as a function of the final sintering temperature. Also shown is the variation in the loss on sintering (LOS) as a function of the final sintering temperature. (b) Diffractograms of samples prepared using the OM and sintered at 500°C, 700°C and 950°C. Illite (Ill); muscovite-2M (Mus); kaolinite (k); anorthite (Anr), albite (Alb), Labradorite (La), and andesine (And); and quartz (Q).

and Brown 1988 reported that in an Anr-Alb-SiO_2-H_2O system, a viscous solution can be readily obtained as the temperature is increased, and it is detected when the quartz diffraction peak begins to disappear; in addition, the plagioclase chemical composition changes very slowly. This dissolution process occurs through a surface-controlled reaction [38] . It is also important to keep in mind that anorthite can be formed at a temperate of approximately 1000°C in a viscous solution of alumina, following the reaction:

$$CaO + Al_2O_3 + SiO_2 \rightarrow CaAl_2Si_2O_8 \text{ or}$$

$$CaO + Al_2Si_2O_7 \rightarrow CaAl_2Si_2O_8 . \qquad (13)$$

However, studies of the formation of anorthite crystalline nuclei and their effects on the increase in strength during sintering are rare in the literature [8] [39] . According to Abdrakhimova and Abdrakhimov, the addition of alkaline oxides to kaolinite and kaolinite-montmorillonite delays the crystallization of cristobalite and lowers the temperature at which the formation of mullite begins [40] . Meanwhile, the eutectic reaction between quartz and albite intensifies the formation of the vitreous phase at low sintering temperatures [41] .

3.4. Prognoses for Molding and/or Extrusion

Figure 6(a) shows the mean granulometry of the LVG and IND materials as well as that of the OM. It also presents an evaluation of the potential uses of the OM in the fabrication of clay products. The IND sample was found to correspond to 8.63% gravel, 79.65% sand, and 11.71% silt plus clay, whereas the LVG sample exhibited a granulometry of 0% gravel, 3.55% sand, and 96.45% slit plus clay. The results for the OM indicated 0% gravel, 45.92% sand, and 54.08% silt plus clay. With this granulometry, and in accordance with the predictions performed with the aid of the diagram of granulometric domains, the OM has the potential for use in the fabrication of clay materials for the construction of hollow products. The characterization of the geotechnical properties of the OM reveals that the OM presented a behavior consistent with that of a low plasticity clay, similar to a sandy or silty clay (PI = 7%; LL = 27% and PL = 20%), whereas the LVG sample exhibited a behavior lying on the boundary between medium and high plasticity (PI = 27%; LL = 50% and PL = 17%) (see Figure 6(b)).

Figure 6(c) presents the workability diagram [42] . In this figure, the prognoses for the molding and/or extrusion of the LVG and OM materials are illustrated in terms of Atterberg limits. From this diagram, it is apparent that the LVG sample exhibited plastic properties that are considered acceptable for molding, whereas the plastic properties of the OM are close to the boundary between acceptable molding and very plastic behavior, with low cohesion. The addition of "sandy" material up to a proportion of 50% in the OM increases the fine sand content and results in a higher content of degreasing material and, consequently, lower plasticity.

3.5. Strength and Durability

Table 3 presents the characterization results for the bricks produced by seven regional artisans and for those fabricated in the laboratory using the OM. In general, clay bricks exhibit several properties that are important for the evaluation of their strength and durability. These properties are related to the quality of the raw material and are directly associated with the manufacturing conditions.

The baking and sintering of clay bricks produce a series of mineralogical, textural, and physical changes that depend on many factors that influence the porosity. The porosity is defined as the ratio of the empty volume in the specimen (pores and cracks) to its total volume. The porosity is an important parameter for clay bricks because of its influence on properties such as chemical reactivity, mechanical strength, durability, and general brick quality. The dimensions and distribution of the pores are influenced by the quality of the raw material, the presence of additives or impurities, the amount of water in the material, and the maximum sintering temperature. The total relative volume of pores that are in contact with the atmosphere, either directly or through other pores

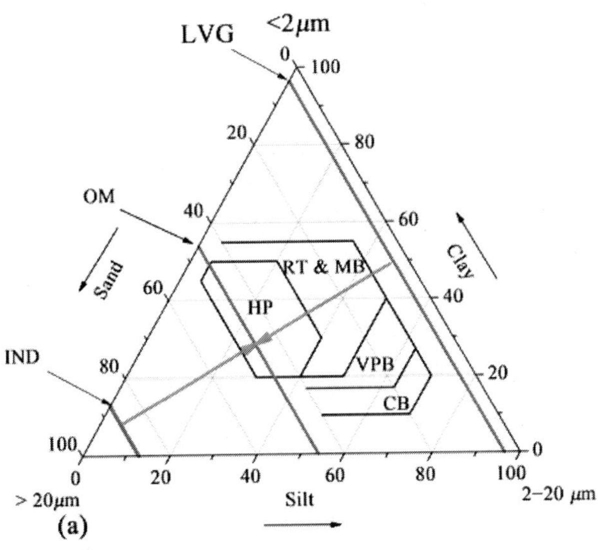

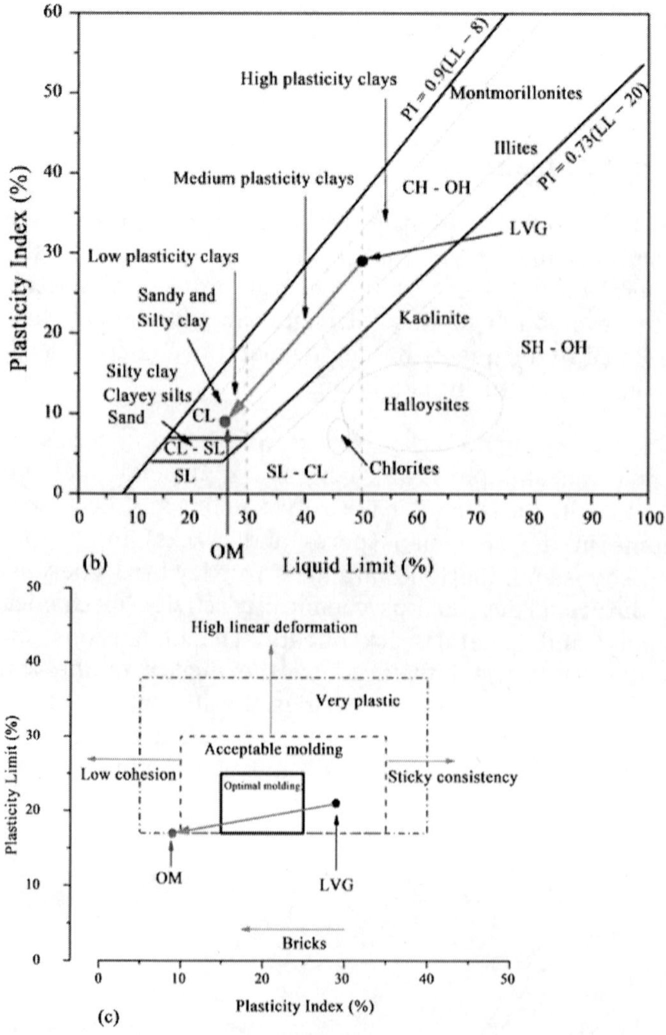

Figure 6. (a) Granulometry of the LVG, IND, and OM samples. The segments proceeding from the sand side to the silt side correspond to the sum of the slit and clay fractions. The segments with arrows point the OM granulometry. The spaces outlined inside the diagram represent the granulometric domains and their corresponding prognoses for use. Conventional brick (CB). Vertical perforated brick (VPB). Roofing tiles and masonry brick (RT & MB). Hallow products (HP). (b) Geotechnical properties of the LVG and OM materials. The two letters used to indicate each subgroup are derived as combinations of the following letters: clay (C), silt (S), organic (O), high plasticity (H), and low plasticity (L). (c) Prognoses for molding and/or extrusion in terms of Atterberg limits.

(see Equation (5)), is represented by the fraction of open pores (f_o) or the apparent porosity. Thc handmade bricks exhibited apparent porosities of between 25 and 34%, whereas the bricks produced in the laboratory exhibited a value of only 21.93%.

The apparent density describes the relationship between the weight of a dry brick and the volume occupied by the brick; thus, it serves as a measure of the proportion of matter that is present in that volume. From this description, it is evident that a higher value of this measure indicates a denser brick, with better mechanical properties and higher durability. Typical values of the apparent density (ρ_b) are between 1.20 and 1.90 g/cm³. The apparent densities of the handmade bricks varied between 1.42 and 1.53 g/cm³, whereas for the bricks produced in the laboratory, this value was 1.85 g/cm³. The solid phase densities (ρ_s) of the handmade bricks varied between 2.02 and 2.27 g/cm³, whereas the value for the bricks fabricated in the laboratory was 2.37 g/cm³.

Pores constitute a large proportion of the volume of a brick, and when the brick is exposed to rain or washing, water penetrates into these pores. The water absorption behavior reflects the brick's water storage capacity and the ability of water to circulate inside the brick, the latter of which can promote deterioration and a reduction in

Table 3. Characterization results for the bricks produced by seven regional producers and the bricks fabricated in the laboratory using the OM.

	P1	P2	P3	P4	P5	P6	P7	Lab
W_1(g) saturated in water	2187	2371	2105	2071	2230	2342	2416	3305
W_2(g) saturated in air	5068	5277	4975	4785	5070	5115	5305	6388
W_3(g) oven-dried	4122	4459	4068	4094	4134	4243	4323	5712
W_4(g) green	-	-	-	-	-	-	-	6091
WA(%)	22.97	18.35	22.31	16.88	22.65	20.53	22.72	11.84
f_o(%) open pores	32.85	28.16	31.62	25.47	32.97	31.42	33.99	21.93
ρ_b(g/cm³)	1.43	1.53	1.42	1.51	1.46	1.53	1.50	1.85
ρ_s(g/cm³)	2.13	2.14	2.07	2.02	2.17	2.23	2.27	2.37
LOS(%)	-	-	-	-	-	-	-	6.64
σ_c(MPa)	5.53	4.78	3.21	6.23	4.22	3.02	5.23	13.89
Std. dev. of σ_c	0.63	0.27	0.71	0.47	1.16	0.43	1.05	0.54
f_{opt}(MPa/kg)	1.34	1.07	0.79	1.52	1.02	0.71	1.21	2.43

*W_1 represents the weight of a saturated brick in water. W_2 represents the weight of a saturated brick in air. W_3 represents the weight of a brick after drying in an oven, and W_4 represents the weight of a green brick. WA represents the relative amount of adsorbed water. f_c is the fraction of space occupied by open pores. ρ_b represents the apparent density, and ρ_s represents the density of the solid phase. LOS indicates the relative mass losses after baking and/or sintering. σ_C represents the compressive strength. Std.dev. of σ_C is the standard deviation of the compressive strength values f_{opt} represents the optimization factor.

*W_1 represents the weight of a saturated brick in water, W_2 represents the weight of a saturated brick in air, W_3 represents the weight of a brick after drying in an oven, and W_4 represents the weight of a green brick. WA represents the relative amount of adsorbed water. f_o is the fraction of space occupied by open pores. ρ_b represents the apparent density, and ρ_s represents the density of the solid phase. LOS indicates the relative mass losses after baking and/or sintering. σ_C represents the compressive strength. Std.dev. of σ_C is the standard deviation of the compressive strength values. f_{opt} represents the optimization factor.

mechanical strength. At temperature below 0°C, the water in the pores freezes, leading to delamination, disintegration, or "cracking". In the presence of soluble salts, the water tends to react with the brick, generating efflorescence. The increase in volume caused by the crystallization of salts can cause severe damage [43] [44] . In Table 3, it is reported that the water absorption (WA) in the handmade bricks varied between 16.88% and 22.97%, whereas that in the bricks fabricated in the laboratory was only 11.84%. The absorption remained below the limit established by the relevant standard; see ASTM standards [3] . However, the absorption of water is an important factor that affects brick durability. Lesser infiltration of water into a brick means that it remains harder and is more resilient to environmental damage. Therefore, the internal structure of a brick should be sufficiently compact to resist the intrusion of water. The nature of the pore network depends not only on the mass loss during baking and sintering but also on the proximity of the particles during the sintering process; closer particles result in a smaller pore size. In general, it is recommended for the water adsorption coefficient to be below 15% [44] [45] .

The compressive strength of a brick provides a measure of its quality. The reported results represent the average values for tests performed on five samples. For the handmade bricks, values ranging from 3 to 5 MPa were recorded, whereas the average value for the brick fabricated in the laboratory was 13.5 MPa; these findings suggest that the bricks fabricated in the laboratory using the OM exhibit a reasonably high compressive strength, corresponding to an improvement of 300% in this parameter. The compressive strength of a material is strongly influenced by the characteristics of the raw materials used and by the production process.

4. CONCLUSIONS

The optimization of the proportions of "clayey" and "sandy" materials used in handmade clay bricks has led to a substantial improvement (300%) in their quality. This improvement should facilitate the more widespread use of these handmade bricks in the local construction industry of southern Ecuador. Clayey materials tend to form solid solutions during baking. The anorthite enrichment of these solid solutions that results from the evolution of andesine contributes to the formation of a stronger matrix among grains of different materials. This can be achieved by using "sandy" materials with high andesine content, which is a common feature of Ecuadorian soil.

The results of IT test performed on disk-shaped samples produced using different mixtures of "clayey" and "sandy" materials enabled the determination of the OM in terms of the value of the optimization factor defined in Equation (10). This is a suitable means of evaluation because the tensile strength primarily depends on the strengths of the mineral grains and of the matrix that binds them together. In addition, there is also a dependence on the chemical composition, the presence of inclusions, and the number and dimensions of the pores. Thus, the IT test has proven to be very sensitive to the strength of the brick binder matrix. The next natural step in this investigation will be to further improve the brick technology; to this end, we are studying the effects of organic additives such as sawdust, rice straw, and cane bagasse on the quality of bricks produced using the OM. It will be necessary to perform additional studies on the effect of anorthite on the increase in strength during sintering and on its behavior in the temperature range between 950°C and 1050°C, in which we can expect to observe the early formation of mullite, among other phenomena.

ACKNOWLEDGEMENTS

The authors wish to express their gratitude to the Prometheus Project of the Secretary of Higher Education, Science, Technology and Innovation of the Republic of Ecuador (Proyecto Prometeo de la Secretaria Superior, Ciencia, Tecnología e Innovación de la República del Ecuador) and to the Technical University of Loja for the support provided during this investigation.

CITE THIS PAPER

Víctor Julio García, Alonso R. Zúñiga-Suárez, Carmen Omaira Márquez,, Juan G. Pérez, Francisco Fernández-Martínez,, Francisco Hernández-Olivares, (2016) Strength Developing in Clay-Andesite Brick. *Materials Sciences and Applications*,**07**,403-420.

REFERENCES

1. Krakowiak, K.J., Lourenco, P.B. and Ulm, F.J. (2011) Multitechnique Investigation of Extruded Clay Brick Microstructure. Journal of the American Ceramic Society, 94, 3012-3022.

2. Monteiro, S.N. and Vieira, C.M.F. (2014) On the Production of Fired Clay Bricks from Waste Materials: A Critical Update. Construction and Building Materials, 68, 599-610.

3. Zhang, L. (2013) Production of Bricks from Waste Materials—A Review. Construction and Building Materials, 47, 643-655.

4. Mukherjee, S. and Ghosh, B. (2013) The Science of Clays: Applications in Industry, Engineering and Environment. Springer, Netherlands.

5. Arsenovic, M., Pezo, L., Stankovic, S. and Radojevic, Z. (2015) Factor Space Differentiation of Brick Clays According to Mineral Content: Prediction of Final Brick Product Quality. Applied Clay Science, 115, 108-114.

6. Wenk, H.R. and Nakajima, Y. (1980) Structure, Formation, and Decomposition of APB's in Calcic Plagioclase. Physics and Chemistry of Minerals, 6, 169-186.

7. Wenk, H.R. (1979) An Albite-Anorthite Assemblage in Low-Grade Amphibolite Facies Rocks. American Mineralogist, 64, 1294.

8. Abdrakhimov, V.Z. (2012) Effect of the Phase Composition on the Durability of Ceramic Facing of the Shakhi-Zinda Ensemble in Samarkand. Glass and Ceramics (English Translation of Steklo i Keramika), 69, 1-3.

9. ASTM D422-63(2007)e2 (2007) Standard Test Method for Particle-Size Analysis of Soils. ASTM International, West Conshohocken, PA.

10. ASTM D4318-10e1 (2010) Standard Test Methods for Liquid Limit, Plastic Limit, and Plasticity Index of Soils. ASTM International, West Conshohocken, PA.

11. ASTM C67-14 (2014) Standard Test Methods for Sampling and Testing Brick and Structural Clay Tile. ASTM International, West Conshohocken, PA.

12. Velasco, P.M., Pilar, M., Ortiz, M., Antonio, M., Giró, M., Melia, D.M., et al. (2015) Development of Sustainable Fired Clay Bricks by Adding Kindling from Vine Shoot: Study of Thermal and Mechanical Properties. Applied Clay Science, 107, 156-164.

13. Hall, C. and Hamilton, A. (2013) Porosity-Density Relations in Stone and Brick Materials. Materials and Structures, 48, 1265-1271.

14. Hangl, M., Borger, A., Danzer, R. and Luxner, H.M. (2002) Application of the Brazilian Disc Test for Strength Measurements on Ceramic Green Bodies. Fracture Mechanics of Ceramics, 13, 159-167.

15. Hungerbühler, D., Steinmann, M., Winkler, W., Seward, D., Egüez, A., Peterson, D.E., et al. (2002) Neogene Stratigraphy and Andean Geodynamics of Southern Ecuador. Earth-Science Reviews, 57, 75-124.

16. Gobierno Autónomo Descentralizado del Cantón Catamayo (2012) Unidad de Gobernabilidad y desarrollo Territorial. Plan de Desarrollo y Ordenamiento Territorial 2014-2022, 654.

17. Kornilov, A.V. and Luzin, V.P. (2004) Effective Methods of Processing Argillaceous Materials for Building Ceramics. Glass and Ceramics, 61, 27-29.

18. Del Coz Díaz, J.J., García Nieto, P.J., álvarez Rabanal, F.P. and Lozano Martínez-Luengas A. (2011) Design and Shape Optimization of a New Type of Hollow Concrete Masonry Block Using the Finite Element Method. Engineering Structures, 33, 1-9.

19. Reed, J.S. (1995) Principles of Ceramics Processing. Wiley-Interscience Publication, Hoboken.

20. Ramachandran, V.S., Paroli, R.M., Beaudoin, J.J. and Delgado, A.H. (2002) Handbook of Thermal Analysis of Construction Materials. William Andrew Publishing/Noyes, Norwich, 491-530.

21. Arsenovic, M., Pezo, L., Mancic, L. and Radojevic, Z. (2014) Thermal and Mineralogical Characterization of Loess Heavy Clays for Potential Use in Brick Industry. Thermochimica Acta, 580, 38-45.

22. Dweck, J. (2008) Qualitative and Quantitative Characterization of Brazilian Natural and Organophilic Clays by Thermal Analysis. Journal of Thermal Analysis and Calorimetry, 92, 129-135.

23. Brindley, W., Maroney, D.M. and Udagawa, S. (1961) High-Temperature Reactions of Clay Mineral Mixture and Their Ceramic Properties: III, Shrinkage and Porosity in Relation to Initial Mineralogy. Journal of the American Ceramic Society, 44, 42-47.

24. Gualtieri, A.F. and Ferrari, S. (2006) Kinetics of Illite Dehydroxylation. Physics and Chemistry of Minerals, 33, 490-501.

25. Chakraborty, A.K. (2014) Phase Transformation of Kaolinite Clay. Springer, New Delhi.

26. Brindley, G. (1961) Effect of Mineralogical Composition of Clays on the Properties of the Fired Body. Journal of the American Ceramic Society, 42-47.

27. Gabor, M., Toth, M., Kristf, J. and Komaromi-hiller, G.I. (1995) Thermal Behavior and Decomposition of Intercalated Kaolinite. Clays and Clay Minerals, 43, 223-228.

28. Ptácek, P., Soukal, F., Opravil, T., Havlica, J. and Brandstetr, J. (2011) The Kinetic Analysis of the Thermal Decomposition of Kaolinite by DTG Technique. Powder Technology, 208, 20-25.

29. Gibbs, H.A.A., O'Garro, L.W. and Newton, A.M. (2000) Differential Thermal Analysis: A Means of Identifying Soil Samples from Barbados, Trinidad, Tobago, St. Vincent, Jamaica and Guyana. Thermochimica Acta, 363, 71-79.

30. MacKenzie, R.C., Ed. (1972) Diferential Thermal Analysis. Academic Press, Lodon, 481-502.

31. Araújo, J.H., De Silva, N.F., Acchar, W. and Gomes, U.U. (2004) Thermal Decomposition of Illite. Materials Research, 7, 359-361.

32. Mazzucato, E., Artioli, G. and Gualtieri, A. (1999) High Temperature Dehydroxylation of Muscovite-2M1: Akinetic Study by in Situ XRPD. Physics and Chemistry of Minerals, 26, 375-381.

33. Guggenheim, S. and Koster van Gross, A.F. (2001) Baseline Studies of the Clay Minerals Society Source Clays: Thermal Analysis. Clays and Clay Minerals, 49, 433-443.

34. Frey, F., Jagodzinski, H., Prandl, W. and Yelon, W.B. (1977) Dynamic Character of the Primitive to Body-Centered Phase Transition in Anorthite. Physics and Chemistry of Minerals, 1, 227-231.

35. Holm, J.L. and Kleppa, O.J. (1968) Thermodynamics of the Disordering Process in Albite. The American Mineralogist. 53, 123-133.

36. Cradarelli, F. (2008) Materials Handbook: A Concise Desktop Reference. 2nd Edition, Springer, Berlin.

37. Fernandes, F.M., Laurenco, P.B. and Castro, F. (2010) Ancient Clay Bricks: Manafacture and Properties. In: Bostenaru, D., Ed., Materials, Technologies and Practice in Historic Heritage Structures, Springer, Berlin, 29-48.

38. Smith, J.V. and Brown, W. (1988) Feldespar Minerals. Springer, Berlin.

39. Abdrakhimov, V.Z. and Abdrakhimova, E.S. (2015) Interrelation of the Phase Composition and Durability of More than 800 Year Old Brick for the Example of the Kazan Kremlin. Glass and Ceramics, 72, 71-75.

40. Abdrakhimova, E.S. and Abdrakhimov, V.Z. (1997) The Influence of Feldspar Concentrate on Phase Transformations in Ceramic Tile. Glass and Ceramics, 54, 326-329.

41. Il'ina, P.V., Skamnitskaya, L.S. and Repnikova, E.A. (1999) Effect of Feldspar Concentrate Composition Mixtures on Structure and Properties of Ceramic. Glass and Ceramics, 56, 259-262.

42. Bain, J.A. and Highley, D.E. (1979) Developments in Sedimentology. Proceedings of the 1978 VI International Clay Conference, Oxford, 10-14 July 1978, 437-446.

43. Alonso-Santurde, R., Coz, A., Viguri, J.R. and Andrés, A. (2012) Recycling of Foundry By-Products in the Ceramic Industry: Green and Core Sand in Clay Bricks. Construction and Building Materials, 27, 97-106.

44. Ajam, L., Ben Ouezdou, M., Felfoul, H.S. and Mensi, R.E. (2009) Characterization of the Tunisian Phosphogypsum and Its Valorization in Clay Bricks. Construction and Building Materials, 23, 3240-3247.

45. Alviset, L. (1987) Matériaux de terre cuite. Techniques de l'ingénieur, C905, 1-20.

12

DEVELOPING SUSTAINABLE HIGH STRENGTH CONCRETE MIXTURES USING LOCAL MATERIALS AND RECYCLED CONCRETE

Anthony Torres*, Alex Burkhart

Department of Engineering Technology, Texas State University, San Marcos, TX, USA

ABSTRACT

This study presents the development of high strength concrete (HSC) that has been made more sustainable by using both local materials from central Texas and recycled concrete aggregate (RCA), which has also been obtained locally. The developed mixtures were proportioned with local constituents to increase the sustainable impact of the material by reducing emissions due to shipping as well as to make HSC more affordable to a wider variety of applications. The specific constituents were: limestone, dolomite, manufactured sand (limestone), locally available Type I/II cement, silica fume, and recycled concrete aggregate, which was obtained from a local recycler which obtains their product from local demolition. Multiple variables were investigated, such as the aggregate type and size, concrete age (7, 14, and 28-days), the curing regimen, and the water- to-cement ratio (w/c) to optimize a HSC mixture that used local materials. This systematic development revealed that heat curing the specimens in a water bath at 50°C (122°F) after demolding and then dry curing at 200°C (392°F) two days before testing with a w/c of 0.28 at 28-days produced the highest compressive strengths. Once an optimum HSC mixture was identified a partial replacement of the coarse aggregate with RCA was completed at 10%, 20%, and 30%. The results showed a loss in compressive strength with an increase in RCA replacement percentages, with the highest strength being approximately 93.0 MPa (13,484 psi) at 28-days for the 10% RCA replacement. The lowest strength obtained from an RCA-HSC mixture was approximately 72.9 (MPa) (10,576 psi) at 7-days. The compressive strengths obtained from the HSC mixtures containing RCA developed in this study are comparable to HSC strengths presented in the literature. Developing this innovative material with local materials and RCA ultimately produces a novel sustainable construction material, reduces the costs, and produces mechanical performance similar to prepackaged, commercially, available construction building materials.

Keywords: High Strength Concrete, Sustainability, Recycled Concrete Aggregate, Local Products, Construction Materials

1. INTRODUCTION

Green construction through sustainable building materials has been an important aspect in the concrete and construction field in the last decade. Using waste products in concrete production is beneficial environmentally and economically: environmentally by replacing a portion of the virgin components with waste materials and environmentally by clean disposal of waste materials. Combining these benefits with locally obtained materials ultimately increases these factors. As Texas is one of the largest producers of cement and aggregate in the nation, it is beneficial to develop a novel sustainable construction building material that utilizes locally available materials. Combining this factor with the reuse of concrete as Recycled Concrete Aggregate (RCA) can drastically increase the sustainable impact of concrete production. Reuse of concrete from demolished structures as aggregates was introduced into practice many years ago, and from the beginning it has been considered in two main environmental aspects: solving the increasing waste storage problem and protection of limited natural sources of aggregates [1] . Presently, most of the demolished concrete extracted from old structures is of relatively good quality [1] . This characteristic demonstrates the potential that recycled concrete aggregate could provide to high strength concrete.

In the past years, improvements have been occurring in concrete and construction material technology. Sustainable use of supplementary materials and revolutionary developments in chemical admixtures has facilitated improvements in the mechanical properties of concrete materials. Such mechanical properties that have drastically been impacted are the strength, density, and the modulus of concrete materials [2] . Through these developments higher strength concrete materials have emerged, known as High Strength Concrete (HSC). However, these chemical and material developments have consistently changed the definition of high strength concrete. The American Concrete Institute (ACI) has defined high-strength concrete as a concrete meeting a high strength that cannot always be achieved routinely, using conventional constituents and normal mixing, placing, and curing practice [2] . In the 1950s, concrete with a compressive strength of 34 MPa (5000 psi) was considered high strength [2] . Today, high-strength concrete is defined as concrete with a specified compressive strength of 55 MPa (8000 psi) or higher [2] . In many markets today, concrete having a specified compressive strength in excess of 69 MPa (10,000 psi) is routinely produced on a daily basis [2] . HSC itself is considered a sustainable construction building material due to the high requirement of silica fume and fly ash, both of which are waste products from other industries, which provides beneficial properties development of concrete [2] . Not only does HSC utilize a high percentage of fly ash and/or silica fume, the high strength requirement results in a higher specific strength of the material. The specific strength of a material is the ratio of the strength to its density. Since the density of HSC doesn't increase significantly to that of conventional concrete, but the strength does, the specific strength is much higher than that of conventional concrete. This aspect impacts sustainable construction through decreased transportation cost and emissions as a smaller structural member can be produced out of HSC, which will require less fuel and produce less carbon emissions during shipping.

In the present research, an attempt has been made to develop HSC mixtures with locally available materials and RCA. The material used in this study included Type I/II Portland cement, silica fume, dolomite, manufactured sand (limestone), and a High Range Water Reducing Agent (HRWRA). Factors such as the aggregate type and size, the curing regimen, the water-to-cement ratio (w/c), and age (7, 14, and 28-day) were investigated.

2. LITERATURE REVIEW

2.1. Background Information

HSC is traditionally composed of cement, coarse aggregate that is much smaller than conventional coarse aggregate ≤9.5-mm (0.375-in.) [2] , fine aggregate, supplemental cementitious materials (SCM) such as silica fume, fly ash, granulated ground blast furnace slag (GGBFS), and quartz powder, fibers, and HRWRA. When used in optimum dosages, the HRWRA reduces the water-to-cement (or water-to-cementitious) ratio while improving the workability (viscosity) of the concrete. The addition of the SCMs enhances the mechanical properties of the cement paste by producing secondary hydrates, filling voids, and enhancing rheology [3] . Due to the burgeoning large structure industry, there are more and more requirements for higher strength concrete. HSC has been in development since the mid 1950's and many researchers are still investigating the optimum and most efficient manner to produce this material [2] . However, modern construction practices are moving towards sustainable construction through lower cost and sustainable materials, therefore, HSC improvements should also be focused on becoming more sustainable and more affordable.

The Environmental Protection Agency (EPA) has estimated that approximately 58% of landfill waste is from construction debris, which a significant portion is concrete and masonry rubble [5] . Recycling waste concrete in the production of new concrete has been investigated since the 1970's [4] -[12] . These studies report a wide variety of results concerning the impact of the RCA on the mechanical and durability properties of the new concrete. The studies have shown that the mechanical properties depend on the properties of the recycled concrete used as well as the amount used. For example, Ravindraraj [9] reported a 9% decrease in compressive strength with a 100% coarse aggregate replacement with RCA, whereas Yamato et al. [10] measured a 45% decrease. Reports have shown the same for the modulus of elasticity in which Gerardu and Hendriks [11] demonstrated a 15% decrease with a 100% replacement, while Frondistou-Yannas [12] reported a 40% decrease at high w/c (0.75) and no decrease at lower w/c (0.55). Due to the wide variation in the properties of concrete with RCA, more research is needed on the local materials used to produce the new and recycled concrete in order to better understand the combined mixture.

2.2. Sustainability

This study focuses on two aspects of sustainable construction building materials; local use of constituent materials and the use of recycled materials. A major concern in the production of HSC is the high cost from shipping such materials as quartz dust, steel or specialty aggregates, and fibers. Most of these constituents are often shipped long distances and internationally in many cases, which increases the cost of the material. It should also be noted that due to the chemical interaction requirements of the silica fume and cement drastically increases the cost of commercially available, prepackaged, HSC products [3] . The commercially available HSC from Ductal uses expensive materials such as ground quartz and fibers that are not traditionally available locally, which increases the cost of the final product. Therefore, the present work focuses on developing HSC mixtures using local materials so that HSC may be made more affordable to wide variety of construction applications. Using local products drastically decreases emissions associated with long shipping routes. The second focus of this study is producing HSC with the inclusion of RCA as partial replacement of virgin coarse aggregate. Replacing virgin materials used in the production of HSC with recycled materials drastically increases the sustainable impact of the material. By using discarded waste material in new construction, the strain for new, virgin, materials is slightly alleviated while also minimizing the demand for landfill space. Additionally, the Leadership in Energy and Environmental Design (LEED) provides a material credit for using building materials or products that have been extracted, harvested or recovered, as well as manufactured, within 500 miles of the project site for a minimum of 10% or 20%, based on cost, of the total material's value [4] . Therefore, using both local materials that make up 100% of the concrete product and using RCA will count double for this LEED requirement.

3. EXPERIMENTAL PROGRAM

3.1. Materials

The final HSC mix design consisted of dolomite coarse aggregate (1.18-mm [0.0469-in.]), manufactured sand, known as "man" sand, which is crushed from limestone with a size of 0.105-mm - 0.60-mm (0.0059-in. - 0.0232-in.), Type I/II cement, silica fume, recycled concrete aggregate with a size of 1.18-mm (0.0469-in.) to match the size of the virgin coarse aggregate, and HRWRA (Master Glenium 3030 from BASF Chemicals). All constituents were obtained from local providers within a 50-mile radius of San Marcos, TX. The aggregate was obtained from local quarries and sieved in the laboratory to achieve a specific size and gradation. The coarse aggregate size was minimized and held at an individual specific size based off the literature [2] [3] [5] -[8] .Table 1 shows the grain size distribution for the fine sand. Table 2 shows the chemical compositions of the Type

Table 1. Grain size distribution for the manufactured sand.

Sieve No.	Sieve size, mm (in.)	Percentage passing
16	1.18 (0.0469)	100
30	0.60 (0.0236)	42.3
50	0.30 (0.0118)	0.0

Table 2. Chemical composition of cement and silica fume.

Compound	Cement	Silica Fume
SiO_2	21.5%	95.8%
Al_2O_3	4.45%	0.18%
Fe_2O_3	3.15%	0.19%
CaO	64.10%	0.30%
K_2O	NA	0.29%
Na_2O	0.52% (Equiv.)	0.20%
MgO	1.90%	0.20%
SO_3	2.89%	0.11%

I/II Portland cement and silica fume. Table 3 shows the physical properties of the coarse aggregate and the RCA. Information about the trial batches can be found in Section 3.2.

3.2. Concrete Mixtures

In order to develop a HSC mixture that used both local materials and RAC, three varying mixtures were developed to first produce a HSC baseline, once an optimum HSC mixture was reached a partial replacement of the coarse aggregate was replaced with RCA, completed in 10%, 20%, and 30% increments. The literature has shown that an upper limit of 30% be used in order to maintain the standard requirement of 5% absorption capacity of aggregates for structural concrete [13] [14] . The literature has also shown that replacement percentages of approximately 15% - 40% begin to diminish the strength of the concrete compared to the control mixture [6] [13] [14] . Therefore, three replacement percentages of 10%, 20%, and 30% were selected for this study. Section 4 describes the results obtained from the trail mixtures and the recommended final RCA-HSC mixture. All trial mixtures were developed based off the recommendations of the literature [1] -[3] [13] [14] and off of data obtained from each trial. The individual trial mixtures were categorized as follows:

Trial A: The mixtures in this category used Type I/II Portland cement, HRWRA, and silica fume. The aggregate for trial group A was Limestone coarse aggregate with a top size of 1.18-mm (0.0469-in.) and manufactured sand with an approximate size range of 0.105-mm - 0.60-mm (0.0059-in. - 0.0232-in.). Three water-to- cement ratios were investigated of 0.32, 0.30, and 0.28.

Trial B: The mixtures in this category used Type I/II Portland cement, HRWRA, and silica fume. The aggregate for trial group B contained Dolomite coarse aggregate with a top size of 2.36-mm (0.093-in.) and Dolomite fine aggregate with an approximate size of 1.18-mm (0.0469-in.). Three water-to-cement ratios were investigated of 0.32, 0.30, and 0.28.

Trial C: The mixtures in this category used Type I/II Portland cement, HRWRA, and silica fume. The aggregate for trial group C consisted of Dolomite coarse aggregate with a top size of 1.18-mm (0.0469-in.) and manufactured sand with an approximate size range of 0.105-mm - 0.60-mm (0.0059-in. - 0.0232-in.). Three water- to-cement ratios were investigated of 0.32, 0.30, and 0.28.

As shown in the three trial mixtures investigated, many of the mixture constituents were held constant in order to determine the impact of the local aggregates and aggregate size. Therefore, HRWRA, silica fume, and Type I/II cement were held constant and weighted based off of the literature. As the literature vastly differs on

Table 3. Physical properties of coarse aggregate and RCA.

Property	Standard	Unit	Limestone Coarse Agg.	Dolomite Coarse Agg.	RCA
Unit Weight	ASTM C29	kg/m³ (lb/ft³)	1442 (90.0)	1859 (116.0)	1411 (88.0)
Water absorption	ASTM C127	%	2.98	3.19	4.12
Bulk Specific Gravity$_{ssd}$[a]	ASTM C127	-	2.57	2.72	2.42
Bulk Specific Gravity$_{od}$[b]	ASTM C127	-	2.51	2.68	2.32

[a]ssd, saturated surface dry condition; [b]od, oven dried condition.

the water-to-cement ratio, three ratios were investigated per trial batch. Due to this change, the HRWRA may vary slightly to maintain a consistent flow-ability/workability of the concrete. As the results will demonstrate, trail C produced the highest results with a w/c of 0.28. Therefore, trial mixture C was used as the baseline HSC for coarse aggregate replacement of RCA. Table 4 displays the mixture proportions for all the trial batches and the RCA-HSC mixtures.

3.3. Specimen Preparation

The aggregate used in this study (coarse and fine) were sieved to obtain the desired size needed as described previously. The aggregates were then thoroughly washed over a No. 200 sieve to remove any fine dust or debris. After washing, the aggregates were oven dried at 44°C (110°F) to achieve a 0% moisture content.

The constituents of each mixture were then mixed for approximately 20 minutes using a laboratory pan mixer. The dry constituents (aggregate, cement, silica fume) were mixed for the first 2 minutes and then 75% of the water was added. After thorough mixing, the HRWRA was added with the remaining 25% of the water. This preparation method was used based off of the literature and experience [1] -[3] [13] -[16] .

3.4. Curing Regimens

In order to minimize as many variables as possible, three curing regimens were investigated on the first mixture developed (Trial A – w/c = 0.32). Trial mixture A was selected to determine the impact of the curing regimen, as it was the first mixture designed and batched. For the first regimen, concrete specimens were cured at room temperature (23°C [73°F]) for the first 24 hours. Once the specimens were demolded, they were moist cured at 23°C (73°F) and a relative humidity of 98% until the day of testing. This curing method is a traditional curing method for conventional concrete as outline in ASTM C192-15 [17] . For reporting purposes this curing method will be reported as Traditionally Cured (TC).

The second curing regimen consisted of curing the specimens at room temperature 23°C (73°F) for the first 24 hours. After the specimens were demolded, the specimens were heat cured in a water bath at a temperature of 50°C (122°F) until the time of testing. For reporting purposes this curing regimen will be reported as Heated Bath Cured (HBC).

The third curing regimen also cured the samples at room temperature 23°C (73°F) for the first 24 hours. After demolding, the specimens were heat cured in a water bath at 50°C (122°F) until 2 days prior to testing. At two days prior to testing, the specimens were removed from the water bath and dry cured at 200°C (392°F). For reporting purposes this curing method will be designated as Oven Dried Curing (ODC). The above curing regimens were developed based on the study by Shaheen (2006) et al. [18] .

3.5. Compression Testing

Compressive strength specimens were molded using 50.8-mm (2-in.) cube molds. Cubes specimens were used to avoid problems with end preparation of cylindrical specimens [2] . After the specimens were properly cured they were individually tested according to BS 12390-3-2009 [19] . The British Standard was used as it provides greater detail to testing hardened concrete cubes in compression than ASTM C 39-15a [19] [20] . An average of three samples were tested per data point reported in the results section. Therefore, a minimum of nine specimens was produced per mixture per w/c in order to obtain a reportable value. Trail mixture A required additional samples as this was the mixture that was used to investigate the curing regimen.

4. RESULTS AND DISCUSSION

4.1. Curing Regimen

The curing regimen results are discussed first as the curing variable was determined prior to curing the majority of the specimens. The first mixture listed in Table 4 (Trial A – w/c = 0.32) was used as the baseline mixture to determine the impact of the curing regimen. Therefore enough specimens were prepared using the batch quantities for this mixture in order to determine the affect of the three curing regimens as outlined above. Figure 1 shows the results of the three curing methods on Trial A – w/c = 0.32).

As seen in Figure 1, the curing regimen significantly influenced the compressive strength of the HSC. The specimens that were cured using the traditional curing method resulted in the lowest average strengths. This is expected as the literature suggests that water curing, such as "full immersion" curing is required for water-to-cement ratios below 0.40 [2] . This is due to the degree of hydration being significantly reduced due to less water in the mixture, therefore it is beneficial to fully immerse the specimens during curing to support additional hydration [2] . Specimens that were cured using the hot bath curing method demonstrated an increase in strength versus the traditional curing method. This curing regimen not only provided a full immersion curing method, but

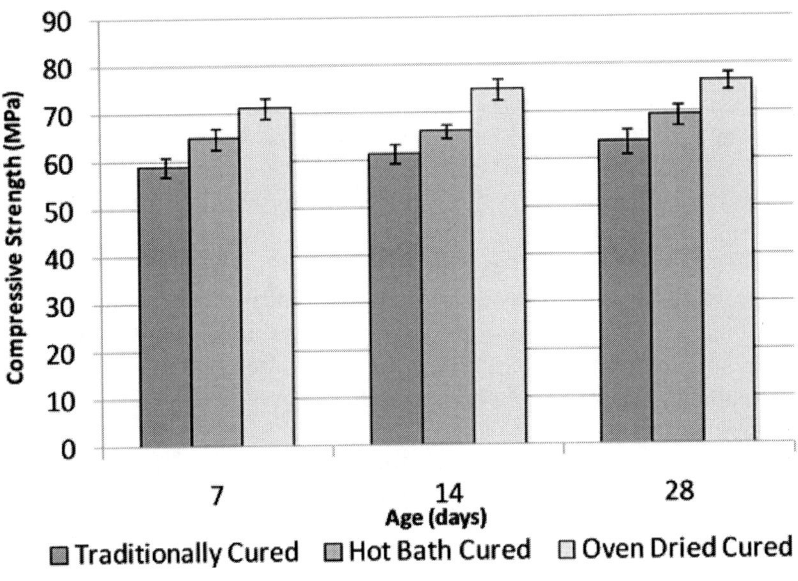

Figure 1. Impact of curing regimen on the compressive strength of Trial A – w/c = 0.32 at varying ages.

Table 4. Mixture proportions of the trail mixtures and RCA-HSC—kg/m³(lb/yd³).

Mixture Name	w/c	Cement	Silica Fume	Coarse Aggregate	Fine Aggregate	RCA	Water	HRWRA
	0.32						166 (280)	16 (27)
Trial A	0.3	520 (876)	100 (69)	825 (1391)	685 (1155)	N/A	156 (263)	18 (30)
	0.28						145 (244)	20 (34)
	0.32						166 (280)	16 (27)
Trial B	0.3	520 (876)	100 (69)	825 (1391)	685 (1155)	N/A	156 (263)	18 (30)
	0.28						145 (244)	20 (34)
	0.32						166 (280)	16 (27)
Trial C	0.3	520 (876)	100 (69)	825 (1391)	685 (1155)	N/A	156 (263)	18 (30)
	0.28						145 (244)	20 (34)
RCA-HSC-10				742.5 (1252)		82.5 (139)		16 (27)
RCA-HSC-20	0.28	520 (876)	100 (69)	660 (1112)	685 (1155)	165 (278)	145 (244)	18 (30)
RCA-HSC-30				577.5 (973)		247.5 (417)		20 (34)

also an increase in curing temperature as the water was held at 50°C (122°F). This form of curing has being used by many authors in order to promote extra hydrates and develop higher early strengths [5] -[8] [12] . This curing regimen resulted in an average 7.3% increase in compressive strength over the traditional curing method. The last curing method, oven dried curing, which utilized a heated water bath and an oven for the last two days prior to testing produced the highest compressive strengths. The oven dried curing resulted in an average 18.8% increase over the traditional curing method and an average 10.5% increase over the hot bath cured. This increase of compressive strength is attributed to acceleration of the hydration reaction when the specimens were kept in the water bath and the formation of secondary calcium silicate hydrate (CSH) from the pozzolanic reaction of the silica fume when the specimens were kept in the oven at 200°C (392°F) for two days prior to testing. These results are consistent with similar results as observed from the literature [5] -[8] [12] . To determine if there is a statistical difference between the three curing regimens a sample t-test was performed. The test was performed with a 95% confidence level and the statistical significance (p value) considered at 0.05 level of confidence was used to analyze the data. All p values recorded were less than 0.05; therefore there is a difference between each curing method. Based off the observation of the curing regimen, all specimens prepared and discussed in the results section have been cured utilizing the oven dried curing regimen.

4.2. Impact of the Water-to-Cement Ratio

The next variable that was investigated in the development of a sustainable HSC mixture was the water-to- cement ratio (w/c). It has been well documented that a lower w/c results in a higher strength concrete and for HSC the w/c should be lower than 0.4 [2] . As previously stated, three w/c ratios were investigated for this study, which are 0.32, 0.30, and 0.28. Figure 2 shows the impact of the w/c on the compressive strength of the specimens from Trial batches A-C.

As shown in Figure 2, the w/c has a significant impact on the strength of the three trial HSC mixtures. Although, the three w/c are only separated by two tenths, the impact is significant in such a controlled mixture as HSC. All three trial batches show that the highest strength occurs at a w/c of 0.28, which is the lowest tested for this study. This is also the case for the age of the HSC. These results are similar to w/c as presented in the literature [5] -[8] [12] . Due to the tight control of these mixtures, the average standard deviation for all three trials at each age was approximately 2.16. A t-test was also completed for the impact of the w/c ratio, which also revealed a difference between each grouping. Therefore, the w/c in fact has an impact on the developed mixtures, as is expected based off the literature and experience. Figure 2 also compares the range of the three trial mixtures; such that Trial C demonstrates the highest recorded compressive strengths. Based off of these results a w/c of 0.28 was selected for the RCA-HSC. These results and results for each trial batch at this w/c are presented in subsequent sections..

4.3. Effect of RCA

The compressive strength was investigated for all trial batches and RCA-HSC specimens produced. To investigate the repeatability of the compressive strengths of these mixtures, testing was conducted on three specimens cast from each of the batches. As previously mentioned the compressive strengths were measured at 7, 14, and 28 days and 50.8-mm (2-in.) cubes were used. To facilitate a better coating of cement paste over the aggregate particles and increase the compressive strength, the dust was removed by washing the aggregates over a No. 200 sieve. The average compressive strengths of the trial mixtures can be seen in Figure 3.

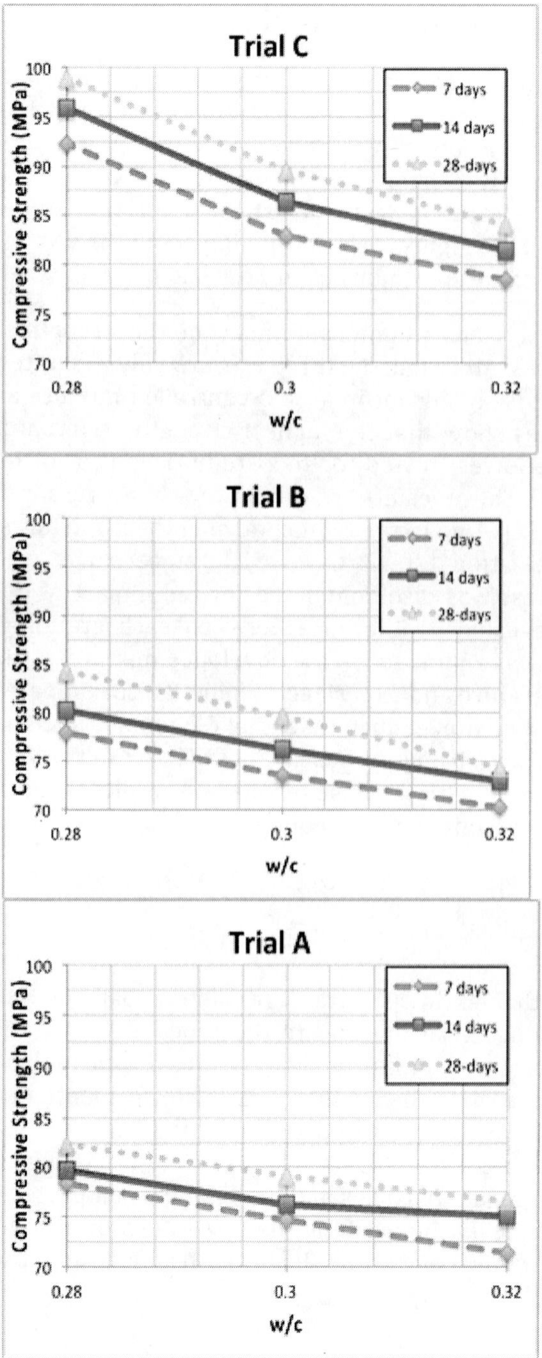

Figure 2. Impact of w/c on the compressive strength of HSC.

Figure 3 demonstrates the impact of age versus the strength of the three trial HSC mixtures. Recall that all of these samples have a consistent w/c of 0.28 and are all cured using the oven dried curing regimen. As expected, the average compressive strength increases in strength as the concrete ages. Additionally, the average standard deviation for this data set was approximately 2.20 due to the tight control of the mixtures. Comparing the three trial mixtures reveals that Trial C produces the highest overall compressive strengths at every age. Trial C has an average percent increase of 25% versus the average compressive strengths of Trial A. Comparing Trials B and A demonstrates an average increase in compressive strength from B to A of only 5%. The increase in strength is due to the size and type of aggregate used. Trial A utilizes limestone coarse aggregate and manufactured sand that is produced from crushed limestone. Limestone is known to be a moderately strong aggregate, but is most commonly used in concrete construction due to it being readily available [21] . Dolomite aggregate was used in Trials B and C and it was selected as it's also readily available in central Texas, but it has a slightly higher density and strength compared to limestone [22] . However, Trial C uses both dolomite as a coarse aggregate and manufactured (limestone) sand as a fine aggregate. This was done due to the higher density of dolomite and the results obtained by Aquino et al. (2010) which shows that fine limestone, such as manufactured sand is known to improve the strength of concrete [21] . Therefore, Trial C combines the impact of both types of locally available aggregates. It is understood that a secondary cause of the minimal increase in strength from trial A to B was due to the size of the aggregates. The smallest aggregate used in Trial B was 1.18-mm (0.0469-in.), which was the largest aggregate size used in both Trials A and C. It should also be noted that all of the strengths at each age and from each trial were higher than 70 MPa (10,150 psi), which is above the range for consideration as HSC. The highest reported strength was from Trial C at a 28-day strength of 98.9 MPa (14,344 psi). Since Trial C had the highest performance, it was selected as the mixture to be used with the RCA replacement. Based off the literature, RCA replacement percentages for coarse aggregates should be less than or equal to 30% replacement by mass, therefore three replacement percentages were selected of 30%, 20%, and 10%. The compressive strength verses age for the RCA-HSC mixtures can be seen in Figure 4.

Figure 4 displays the compressive strength results for the developed sustainable high strength concrete using RCA and local materials. Trial C compressive strength data was included in this plot for comparison as this mixture was the baseline for all RCA-HSC mixtures. The results ultimately show a decrease in strength by including any percent replacement percentage investigated in this study. The decrease in strength increased with an increase in percentage of RCA used, i.e. the highest strength loss came from the 30% replacement of RCA coarse aggregate. The average strength loss was 7.9%, 14.4%, and 22.6% for the 10%, 20%, and 30% RCA inclusion respectively and an average standard deviation for each grouping of 2.63. This decrease in compressive strength is typical for most concrete mixtures containing RCA, as the quality of the RCA aggregate is significantly lower than virgin aggregates [5] -[8] [12] . The major issue when dealing with RCA is the fact that the aggregate type and condition is unknown. Additionally, it is often difficult to remove debris or foreign matter that could be embedded in the RCA that could affect the performance of the new concrete. Although, there is a minor reduction in strength when using RCA as partial coarse aggregate replacement in HSC, it should be noted that the lowest recorded strength from RCA-HSC-30 was 79.1 MPa (11,476 psi) at 28-days. This strength is still high enough to be

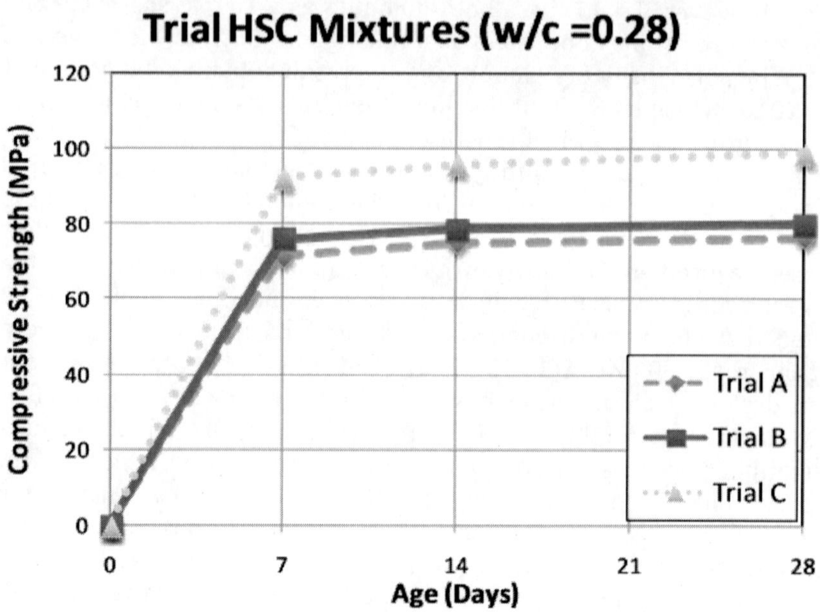

Figure 3. Average compressive strength of the trial HSC mixtures.

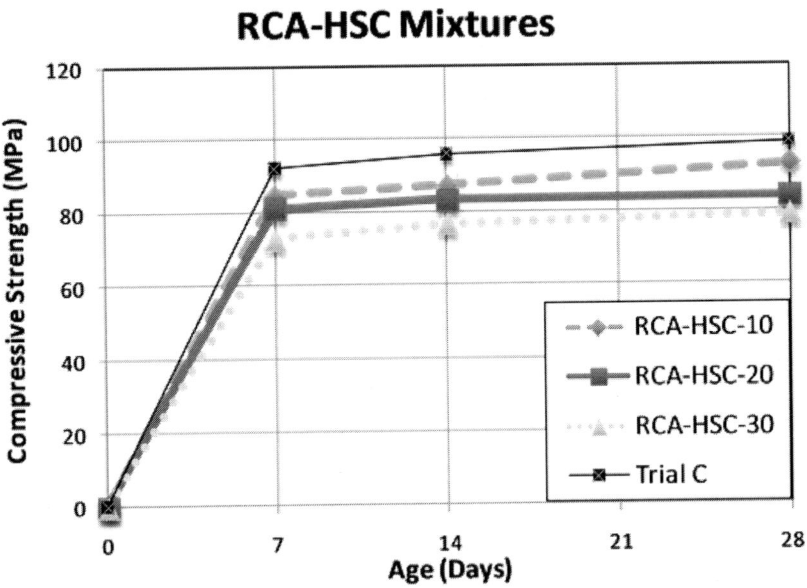

Figure 4. Average compressive strength of the RCA-HSC mixtures.

considered HSC and is also approximately the same strength at 28-days of the Trial B mixture that contained no RCA elements. A drop in strength was expected based off the literature; therefore the highest performing trial mixture was selected such that the novel sustainable construction building material would still have a high strength.

5. CONCLUSION

A novel sustainable construction building material has been developed that has high strength, uses recycled elements, and all the constituents were obtained locally. This study focused on developing a novel construction building material that can impact the sustainable construction building movement by not only developing a HSC, but the HSC has been made more sustainable by utilizing RCA and all local constituents. This study focused on maintaining consistent variables in order to assess the impact of the RCA on the HSC. This was developed by first assessing the impact of the curing regime on the strength of the concrete. Three regimens were investigated, which were cultivated based off the literature. Once an optimum curing regimen was reached, the w/c ratio was investigated. Again three different w/c ratios were investigated, which were also selected based off the literature. After the optimum curing regimen and w/c ratio was obtained, three different mixture designs were developed and compared. The three mixture designs consisted of two different locally available aggregate types and the size of the aggregate. The compressive strength measurements were compared at three testing ages of 7, 14, and 28 days. The highest performing HSC mixture was then used to investigate the impact of the RCA. RCA was used to partially replace the coarse aggregate in the top performing trial mixture at 10%, 20% and 30% replacement by mass. The results show that the RCA-HSC had a low end strength of 72.9 MPa (10,576 psi) at 7 days from the 30% RCA replacement and a high end strength of 93.0 MPa (13,484 psi) at 28 days from the 10% RCA replacement. Therefore, a novel sustainable HSC mixture has been produced that utilizes 100% local materials.

REFERENCES

1. Andrzej, A and Alina, K (2002) Influence of Recycled Aggregates on Mechanical Properties of HS/HPC. Cement and Concrete Composites, 24, 269-279.

2. ACI Committee 363 (2010) Report on High-Strength Concrete. ACI 363R-10, American Concrete Institute Committee 363, Farmington Hills, MI.[Paper reference:13]

3. Dili, A.S. and Santhanam, M. (2004) Investigations on Reactive Powder Concrete: A Developing Ultra High Strength Technology. The Indian Concrete Journal, 74, 33-38. [Paper reference:5]

4. Azhar, S., Carlton, W.A., Olsen, D. and Ahmad, I. (2011) Building Information Modeling for Sustainable Design and LEED® Rating Analysis. Automation in Construction, 20, 217-224.

5. Rahal, K. (2007) Mechanical Properties of Concrete with Recycled Coarse Aggregate. Building and Environment, 42, 407-415.

6. Buck, A.D. (1977) Recycled Concrete as a Source of Aggregate. ACI Journal, 74, 212-219.

7. Hansen, T.C. and Narud, H. (1983) Strength of Recycled Concrete Made from Crushed Concrete Coarse Aggregate. Concrete International, 1, 79-83.

8. Forster, S.W. (1986) Recycled Concrete as Aggregate. Concrete International, 8, 34-40.[Paper reference:5]

9. Ravindraraj, R.S., Steward, M. and Greco, D. (2000) Variability of Recycled Concrete Aggregate and Its Effect on Concrete Properties—A Case Study in Australia. International Workshop on Recycled Concrete, JSPS 76 Committee on Construction Materials, Tokyo, September 2000, 27-42.

10. Yamato, T., Emoto, Y. and Soeda, M. (1998) Mechanical Properties, Drying Shrinkage and Resistance to Freezing and Thawing of Concrete Using Recycled Aggregate. ACI Special Publication SP 179-7, American Concrete Institute, Farmington Hills, MI, 105-121.

11. Gerardu, J.J.A. and Hendriks, D.F. (1985) Recycling of Road Pavement Materials in the Netherlands. Rijkswaterstaat Communications No. 38, The Hague.

12. Frondistou-Yannas, S. (1977) Waste Concrete as Aggregate for New Concrete. ACI Journal, 78, 373-376.[Paper reference:6]

13. Ajdukiewicz, A. and Kliszczewicz, A. (2002) Influence of Recycled Aggregates on Mechanical Properties of HS/HPC. Cement and Concrete Composites, 24, 269-279.

14. Tu, T.-Y., Chen, Y.-Y. and Hwang, C.-L. (2006) Properties of HPC with Recycled Aggregates. Cement and Concrete Research, 36, 943-950.

15. Rao, A., Jha, K.N. and Misra, S. (2007) Use of Aggregates from Recycled Construction and Demolition Waste in Concrete. Resources, Conservation and Recycling, 50, 71-81.

16. Evangelista, L. and de Brito, J. (2007) Mechanical Behaviour of Concrete Made with Fine Recycled Concrete Aggregates. Cement and Concrete Composites, 29, 397-401.

17. ASTM Standard C192-15 (2015) Standard Practice for Making and Curing Concrete Test Specimens in the Laboratory. ASTM International, West Conshohocken, PA.www.astm.org

18. Shaheen, E. and Shrive, N.J. (2006) Optimization of Mechanical Properties and Durability of Reactive Powder Concrete. ACI Materials Journal, 103, 444-451.

19. BS EN 12390-3:2009 (2009) Testing Hardened Concrete Compressive Strength of Test Specimens B/517/1, 22.

20. ASTM Standard C 39-15a (2015) Standard Test Method for Compressive Strength of Cylindrical Concrete Specimens. ASTM International, West Conshohocken, PA.www.astm.org

21. Aquino, C., Inoue, M., Miura, H., Mizuta, M. and Okamoto, T. (2010) The Effects of Limestone Aggregate on Concrete Properties. Construction and Building Materials, 24, 2363-2368.

22. Mikhailova, O., Yakovlev, G., Maeva, I. and Senkov, S. (2013) Effect of Dolomite Limestone Powder on the Compressive Strength of Concrete. Procedia Engineering, 57, 775-780.

13

KINETIC MODEL OF NON-ISOTHERMAL CRYSTAL NUCLEATION WITH TRANSIENT AND ATHERMAL EFFECTS

Andrzej Ziabicki[1], Beata Misztal-Faraj[1], and Leszek Jarecki[1]

[1]Institute of Fundamental Technological Research, Polish Academy of Sciences, Pawinskiego 5 B, Warsaw, Poland

ABSTRACT

A kinetic model of primary homogeneous non-isothermal crystal nucleation with transient and athermal effects is developed. For comparison, steady-state and transient isothermal nucleation rates are considered. Kinetic equation for the development of cluster size distribution provides the basis for the model. Transient effects are characterized by the longest relaxation time which increases with temperature at low and moderate undercooling. In isothermal conditions, nucleation rate is controlled by thermal mechanism; in non-isothermal conditions, there appears also athermal mechanism. Closed-form analytical formula for the development of transient cluster size distribution in single-relaxation-time approximation is derived for non-isothermal processes, as well as thermal and athermal nucleation rates and total number of nuclei produced in a cooling or heating run. The transient term contributes to isothermal nucleation kinetics the more the higher is temperature. Under non-isothermal conditions, the relaxation time contributes to the nucleation kinetics by the product with the cooling/heating rate. Considerable transient effects should be expected for the relaxation times as long as 10^2–10^5 s. Contribution of thermal nucleation to the concentration of nuclei is inversely proportional to the temperature rate, while the contribution of athermal nucleation depends on the temperature interval of cooling or heating. Our kinetic model indicates similarities in the nucleation mechanisms in polymers and metals undergoing crystallization. Example computations are presented for molten indium and a linear polymer—polyhydroxybutyrate (PHB). A low-temperature limit is predicted for the nucleation mechanism in PHB, while for indium the mechanism is active in the entire temperature range.

INTRODUCTION

Crystallization plays important role in material science and technology. Crystal structure, degree of crystallinity, crystal size distribution, and molecular orientation strongly affect mechanical, sorptional, and optical properties of many metals, ceramics, and polymers. Kinetics of the nucleation-controlled crystallization are sensitive to temperature, to temperature history, and to the rate with which the temperature changes during the crystallization time. Optimum temperature and cooling/heating rates must be known whenever desired crystallinity of the material is sought. In variable temperature, nucleation involves different mechanisms responsible for steady-state, transient, and athermal (memory) effects. Kinetics of nucleation have been widely described in books and review articles [1–7]. Whereas steady, isothermal nucleation has been reasonably well studied, the mechanisms involved in variable external conditions (temperature, stress, etc.) still require theoretical and experimental research.

Crystallization involves two successive processes: formation of primary nuclei and crystal growth. Primary nucleation initiates most of the solidification processes in liquids and in solid–solid transformations. The basis of primary crystal nucleation is the formation of atomic or molecular clusters which grow spontaneously to macroscopic dimensions. The nuclei can be formed homogeneously from a pure amorphous phase or by heterogeneous mechanism on foreign particles. The knowledge on the role of the state parameters (temperature, concentration, pressure, stresses) in the nucleation kinetics is important for controlling structure formation in low and high molecular materials. In the early approaches, the thermodynamic criterion of phase instability proposed by Gibbs was used for developing kinetic theory of nucleation in vapors, gasses, and condensed matter [1, 8–12].

In the case of isothermal conditions, the kinetic theories concern steady-state and non-steady-state nucleation. Volmer and Weber [8] originally developed the theory for the condensation of saturated vapors as a process activated by thermal fluctuations and controlled by free energy barrier. The formula for steady-state homogeneous nucleation rate at constant temperature in such systems was derived by Becker and Döring [10], Zeldovich [11], and Frenkel [1] basing on the steady-state cluster volume distribution. With the theory of absolute reaction rates proposed by Eyring [13], homogeneous nucleation rate for condensed systems at constant temperature was derived by Turnbull and Fisher [12]. Spherical kinetic units (atoms) subjected to aggregation into spherical clusters were considered in that model in application to molten metals.

In the classical models of nucleation in condensed phases, various shapes of the aggregating units, nuclei, and clusters were assumed (spheres, cylinders, polyhedra, etc.). Asymmetric kinetic units forming cylindrical clusters [14] or rectangular parallelepipeds [15] were considered in the kinetic theory applied to polymers. For non-spherical clusters, a formal treatment covering cluster volume and shape distribution was proposed by Ziabicki [16]. The kinetic approach to steady-state homogeneous nucleation rate in polymers was verified experimentally by Turnbull et al. [17, 18] and for liquid metals [19, 20]. In considering kinetics of isothermal nucleation in condensed matter, it is important to account for transient behavior. Kinetic theory of non-steady-state isothermal nucleation controlled by transient distribution of cluster size was considered analytically by Zeldovich [11], Collins [21], Kashchiev [22], Chakraverty [23], Wakeshima [24], Feder [25], Andres [26], and by computer simulations by Kelton [27]. These theories account for a transient time during which the rate of nuclei production approaches steady-state level. In metallic and silicate glasses, the transient time is long enough to be observed experimentally due to the incubation periods representing the time taken for a concentration of nuclei to be achieved [28–31].

The results of computer simulation on transient isothermal nucleation rate and temperature dependence of the time lag in one-component glasses [27] are in best agreement with the Kashchiev theory [22]. The computed time lag is nearly equal to the longest relaxation time of the cluster size distribution, as predicted in [22]. Increasing of the temperature results in strong decrease in the computed time lag from several tens of hours to milliseconds in the temperature range of ±100 K around the maximum nucleation rate. Basing on the experimental results of Kalinina [32], it was shown by Kelton and Greer [33] that transient isothermal nucleation in lithium disilicate glasses proceeds accordingly to time evolution of cluster distribution from a predetermined initial state, as predicted by the classical theory.

Using computer simulation, Kelton and Greer [34–36] have predicted that the transient phenomena are important also in non-isothermal homogeneous nucleation (alkali silicates, metallic glasses). The authors have shown that the transient effects reduce nucleation rate the more so, the higher is cooling rate. The transient effects reduce critical cooling rate for achieving glassy state by two to three orders of magnitude, relative to steady-state kinetics, in agreement with the experimental data. Transient effects in non-isothermal nucleation play significant role in the glassy state [31, 37].

For crystallization from metallic liquids, transient effects in nucleation kinetics can play a prominent role at extremely rapid cooling [7] due to very short time lags (relaxation times) in the molten state. At such fast cooling processes, deviation from the steady-state cluster size distribution should be accounted for. Formation of metallic glasses provides an example. It has been shown experimentally [34] that deviation from the steady-state results in reduction in the number of nuclei produced during the quench, scaled with the cooling rate lower by one to tree orders of magnitude than predicted for a steady-state process.

When chemical composition of the nucleating phase differs from that of the mother phase, the effects of long-range diffusion coupled with the attachment–detachment rates are important in the nucleation kinetics. In the case of one-component systems and significantly asymmetric kinetic units, long-range translational diffusion is coupled with rotational diffusion [38, 39] and may affect transient behavior of isothermal and non-isothermal nucleation.

In the present paper, we derive a simple kinetic model of primary, homogeneous, non-isothermal nucleation. We modify the transient kinetic theory of isothermal nucleation proposed by Kashchiev [22] to obtain a closed-form, analytical expression for the kinetics of transient nucleation in one-component systems under non-isothermal conditions. In our model, the simplest cubic symmetry of the kinetic elements and the longest relaxation time are assumed which allows to analyze the effects considered in this paper. Orientation effects involving non-isodiametric symmetry will be subject of another paper.

The transient non-isothermal nucleation rate is based on time-dependent cluster size distribution controlled by the thermodynamic factors and the longest temperature-dependent relaxation time. Transient effects with temperature-dependent relaxation time are present also in isothermal conditions. That is why relaxation time appears in the non-isothermal model. Contribution of athermal nucleation in the kinetic model is also accounted for. The concept of athermal nucleation was introduced by Fisher et al. [40], Ziabicki [41], and Kashchiev [5].

We analyze individual factors and mechanisms involved in nucleation kinetics and indicate those processing conditions and material parameters which play decisive role in the overall nucleation rate, in the reference to steady-state and transient isothermal processes. Using our model, numerical evaluation of nucleation kinetics is performed for two example materials: metallic indium and the linear polymer polyhydroxybutyrate (PHB).

Cluster size distribution and nucleation rate

To describe the formation of atomic or molecular clusters as the basis of crystal nucleation, we follow the classical approach by Volmer and Weber [8], Becker and Döring [10], Turnbull and Fisher [12] and formulate continuous distribution of cluster density ρ(v)ρ(v) in the space of their volumes *v*

$$\begin{aligned} \rho(v) &= \frac{1}{V}\frac{\mathrm{d}N(v)}{\mathrm{d}v} \\ \rho(v = v_0) &= \rho_0 \\ \rho_0 v_0 &= \frac{N_1}{V} \end{aligned} \tag{1}$$

where v_0 is the volume of a single kinetic unit undergoing aggregation in the system of volume *V*. For condensed systems, we have ρ0=1/v20ρ0=1/v02.

In our approach with the cluster volume as the size variable, *v*, the continuity equation of the classical kinetic theory of nucleation for cluster size distribution reads

$$\begin{aligned} &\frac{\partial \rho(v,t)}{\partial t}+\frac{\partial j(v,t)}{\partial v}=0 \\ &j(v,t)=-D(v,t)\left(\frac{\partial \rho}{\partial v}+\frac{\rho}{kT}\frac{\partial \Delta G(v,t)}{\partial v}\right) \end{aligned} \tag{2}$$

Equation (2) describes kinetics of quasi-chemical bimolecular reactions between clusters, $v > v_0$, and single kinetic units, $v = v_0$. D(v,t)D(v,t) is a "diffusion coefficient" of clusters in the space of their volumes v, and ΔG(v,t)ΔG(v,t) is free enthalpy of formation a v-size cluster from the equivalent volume of single elements. The assumed isodiametric geometry makes volume of the molecular clusters, v, a single variable of the model. With this variable, basic non-isothermal and transient effects are treated. For cubic kinetic elements assumed in the model, we have $v_0 = d_0^3;\ v = d^3$. Concentration of single kinetic units (atoms, molecules, molecular segments) is assumed to be large, compared to that of dimers, trimers, and higher-order clusters.

Free enthalpy of cluster formation, ΔG, includes a bulk term, Δg, proportional to undercooling Tm−TTm−T and a positive term characterized by surface free energy density σ

$$\Delta G(v,T)=\Delta g v+6\sigma v^{2/3}=-\Delta h\frac{(T_m-T)}{T_m}v+6\sigma v^{2/3} \tag{3}$$

Δh is the heat of melting and T_m is the equilibrium melting temperature.

Differentiation of ΔG with respect to v yields critical cluster size v^* and the maximum free energy ΔG^* providing thermodynamic barrier for nucleation

$$\frac{d\Delta G}{dv}=0 \Rightarrow \begin{cases} v^*(T)=\dfrac{64\sigma^3}{|\Delta g|^3}=\dfrac{64\sigma^3 T_m^3}{\Delta h^3 (T_m-T)^3} \\ \Delta G^*(T)=\dfrac{32\sigma^3}{\Delta g^2}=\dfrac{32\sigma^3 T_m^2}{\Delta h^2 (T_m-T)^2} \end{cases} \tag{4}$$

Maximum free energy ΔG∗ΔG∗ provides potential barrier for the flux of clusters in the space of their volumes v. Thermal nucleation rate N˙N˙ is defined as the number of overcritical clusters (nuclei) produced in unit time and unit volume of the system

$$\dot{N}(T) = j(v*, T) = -D(v*)\left(\frac{\partial \rho}{\partial v}\right)_{v=v*} \tag{5}$$

The range of temperatures in which homogeneous thermal nucleation exists is limited. The upper limit is defined by equilibrium melting temperature T_m

$$T = T_m \Rightarrow \begin{cases} v^*(T) \to \infty \\ \Delta G^*(T) \to \infty \\ \dot{N}(T) = 0 \end{cases} \tag{6}$$

With increasing undercooling, critical cluster size v^* decreases. At the lower limiting temperature T_{lim}, v^* drops down to the level of a single kinetic unit

$$\begin{gathered} T = T_{\lim} \Leftrightarrow v* = v_0 \\ T_{\lim} = T_m\left(1 - \frac{4\sigma}{\Delta h v_0^{1/3}}\right) \end{gathered} \tag{7}$$

With Eq. (7), we introduce the low-temperature bound T_{lim} for crystallization controlled by homogeneous primary nucleation. Below T_{lim}, all clusters present in the system and single elements behave like effective nuclei and are able to grow to macroscopic dimensions.

Such a behavior has been analyzed by George [42] as a "nucleative collapse" in the case of high-speed melt spinning. When temperature T_{lim} is reached, all clusters including monomers become effective nuclei and the kinetics of crystallization is solely controlled by diffusional growth. Evaluation of Eq. (7) using material data from Table 1 yields for PHB T_{lim} = 359 K while for indium, T_{lim} calculated from Eq. (7) is negative. Consequently, primary nucleation mechanism in indium is active in the entire range of temperatures. Non-nucleative temperature range seems to be involved in materials undergoing spinodal decomposition [43, 44] where the transition rates are controlled by diffusional growth. Such a behavior may also appear in crystallization of highly oriented polymers [42, 45] where orientation entropy strongly reduces the critical cluster size.

Table 1. Material characteristics of indium and polyhydroxybutyrate (PHB)

Material characteristics		Indium		PHB	
T_m (K)	Melting temperature	429.80	[54]	470.0	[55]
Δh (J/cm^3)	Heat of fusion per unit volume	209.20	Calculated from [54, 56]	185.0	Calculated from [55]
$\Delta h/\gamma$ (J/g)	Heat of fusion per unit mass	28.62	[54]	146.0	[55]
σ (J/cm^2)	Interface free energy density	2.730×10^{-6}	Calculated from [7]	1.458×10^{-6}	Calculated from [50]
E_D (kJ/mole)	Activation energy	4.39	Calculated from [57]	24.36	[58]
d_0 (cm)	Average dimension of kinetic element	2.97×10^{-8}	Calculated from [56]	1.33×10^{-7}	Calculated from [55, 59]
v_0 (cm^3)	Volume of kinetic element	2.62×10^{-23}	[56]	2.35×10^{-21}	[55, 59]
γ (g/cm^3)	Crystal density	7.31	[56]	1.26	[55]

Above the critical transition temperature, $T > T_m$, the flux of cluster growth disappears and Eq. (2) reduces to

$$j_{eq}(v) = 0$$
$$\frac{d\rho_{eq}}{dv} + \frac{\rho_{eq}}{kT}\frac{d\Delta G_{eq}}{dv} = 0 \quad (8)$$

and cluster distribution assumes the Boltzmann form

$$\rho_{eq}(v,T) = \rho_0 \exp\left[\frac{-\Delta G_{eq}(v,T)}{kT}\right]$$
$$= \rho_0 \exp\left[\frac{\Delta hv(T_m - T)/T_m - 6\sigma v^{2/3}}{kT}\right] \quad (9)$$

Asymptotically, we have

$$\Delta G_{eq}(v) = \begin{cases} 0 & \text{for } v = v_0 \\ \infty & \text{for } v \to \infty \end{cases} \quad (10)$$

Steady-state isothermal nucleation rate

To describe steady flow of single kinetic units into the system of clusters and steady flow of large clusters leaving the system, we follow principles of the classical nucleation theory [8, 10, 12]. The boundary conditions in the space of cluster volumes, v, include constant concentration of single elements and zero concentration of very large clusters

$$\begin{aligned} &\rho_{st}(v = v_0) = \rho_0 = \text{const} \\ &\rho_{st}(v \geq v_{\max}) = 0 \\ &v_{\max} \cong 2v* \end{aligned} \tag{11}$$

The steady-state model involves an absorbing boundary at $v = v_{max}$ and a providing boundary at$v = v_0$. Without an excessive error, the upper limit v_{max} can be approximated with double critical size.

In steady-state isothermal conditions, Eq. (2) reduces to

$$\begin{aligned} &\frac{\partial \rho_{st}}{\partial t} = 0 \\ &j_{st}(v) = \text{const} \\ &\frac{d\rho_{st}(v,T)}{dv} + \frac{\rho_{st}(v,T)}{kT}\frac{d\Delta G(v,T)}{dv} = \frac{\text{const}(T)}{D(v,T)} \end{aligned} \tag{12}$$

Exact solution of Eq. (12) can be obtained in the form of integrals. In the space of cluster volume, we use, however, a more handy analytical approximation proposed by Zeldovich [11] and Frenkel [1]

$$\begin{aligned} &\rho_{st}(v,T) = \frac{1}{2}\rho_0 e^{\frac{-\Delta G(v)}{kT}}\{1 - erf[Z(v - v*)]\} \\ &\frac{d\rho_{st}}{dv}(v,T) = -\rho_0 e^{\frac{-\Delta G(v)}{kT}} \\ &\times\left[\{1 - erf[Z(v - v*)]\}\frac{1}{2kT}\frac{d\Delta G}{dv} + \frac{Z}{\sqrt{\pi}}e^{-Z^2(v-v*)^2}\right] \end{aligned} \tag{13}$$

where Z is the half-width of the potential function $\Delta G(v)$, known as the Zeldovich factor

$$Z(T) = \sqrt{\left.\frac{-1}{2kT}\frac{d^2\Delta G}{dv^2}\right|_{v*}} = \frac{\Delta h^2(T_m - T)^2}{8\sqrt{6kT}\sigma^{3/2}T_m^2} \tag{14}$$

In the critical point, $v = v^*$, steady-state distribution of clusters in the space of cluster volume reduces to

$$\begin{aligned}
\rho_{st}(v*) &= \frac{1}{2}\rho_0 \exp\left[\frac{-32\sigma^3 T_m^2}{kT\Delta h^2 (T_m - T)^2}\right] \\
\left.\frac{d\rho_{st}}{dv}\right|_{v*} &= \frac{-Z}{\sqrt{\pi}}\rho_0 \exp\left(\frac{-\Delta G^*}{kT}\right) \\
&= \frac{-\Delta h^2 (T_m - T)^2}{8\sqrt{6\pi kT}\sigma^{3/2} T_m^2}\rho_0 \exp\left[\frac{-32\sigma^3 T_m^2}{kT\Delta h^2 (T_m - T)^2}\right]
\end{aligned} \tag{15}$$

The “diffusion coefficient” in the space of cluster volumes, $D(v)$, is proportional to the cluster surface area, $v^{2/3}$, and to the frequency of the molecular motions. For the “diffusion coefficient,” we assume the Arrhenius approximation with constant transport barrier E_D, and in case of cubic clusters, we obtain

$$D(v, T) = 3v^{2/3} v_0^{4/3} \frac{kT}{h} \exp\left(\frac{-E_D}{kT}\right) \tag{16}$$

which at v=v∗(T)v=v∗(T) reduces to

$$D[v^*(T), T] = \frac{48\sigma^2 v_0^{4/3} T_m^2}{\Delta h^2 (T_m - T)^2} \frac{kT}{h} \exp\left(\frac{-E_D}{kT}\right) \tag{17}$$

Thus, we obtain steady-state nucleation rate in the form

$$\begin{aligned}
\dot{N}_{st}(T) &= -D(v^*) \left.\frac{d\rho_{st}}{dv}\right|_{v=v^*} \\
&= I_0 \exp\left(\frac{-E_D}{kT}\right) \exp\left[\frac{-32\sigma^3 T_m^2}{kT\Delta h^2 (T_m - T)^2}\right] \\
I_0 &= v_0^{-2/3} \left(\frac{6\sigma kT}{\pi h^2}\right)^{1/2}
\end{aligned} \tag{18}$$

Transient isothermal nucleation rate

In any real system, the distribution of clusters is subject to molecular rearrangement, and steady state is reached only asymptotically. Transient isothermal solutions of the continuity equation, Eq. (2), have been obtained by Collins [21], Kashchiev [22], Chakraverty [23], and others. For the cluster size distribution in the space of their volumes, separation of variables in Eq. (2) yields an infinite sum of exponentials

$$\rho(v,t) = a_0(v) + \sum_{i=1}^{\infty} a_i(v)\exp(-t/\tau_i) \tag{19}$$

where τiτi denotes relaxation times and a_i are v-dependent parameters.

We approximate the distribution, Eq. (19), by a single time-dependent term with the longest relaxation time τ=τ1τ=τ1

$$\rho(v,t) \cong a_0(v) + a_1(v)\exp(-t/\tau) \tag{20}$$

With the boundary conditions

$$\begin{aligned} \rho(v,t=0) &= \rho_{in}(v) \\ \rho(v,t\to\infty) &= \rho_{st}(v) \end{aligned} \tag{21}$$

we obtain the following distribution controlled by single-relaxation time

$$\rho(v,t) \cong \rho_{\mathrm{st}}(v) + [\rho_{\mathrm{in}}(v) - \rho_{\mathrm{st}}(v)]\exp(-t/\tau) \tag{22}$$

where τ is a temperature-dependent material characteristics. The nucleation time lag θθ is expressed by the longest relaxation time τ by the formula θ=π2τ/6θ=π2τ/6 [22]. It is evident that at infinitely long τ, the initial distribution ρin(v)ρin(v) remains unchanged, and at $\tau = 0$, steady state is reached instantaneously.

By differentiation of Eq. (22), we obtain the following equation for the rate of variation in the distribution in the processes controlled by the longest relaxation time

$$\frac{\partial \rho(v,t)}{\partial t} = \frac{\rho_{\mathrm{st}}(v) - \rho(v,t)}{\tau} \tag{23}$$

The rate of variation is proportional to the difference between the asymptotic steady-state distribution and the instantaneous distribution, as well as inversely proportional to the temperature-dependent relaxation time τ. This relationship will be used below in this paper for determining the development of cluster size distribution in non-isothermal processes, by integrating Eq. (23) with τ and ρst(v)ρst(v) dependent on varying temperature.

Application of Eqs. (5), (18), and (22) yields isothermal transient nucleation rate

$$\dot{N}_{\mathrm{trans}}(t) = -D(v*)\left[\left.\frac{\partial \rho_{st}}{\partial v}\right|_{v*} + \left(\left.\frac{\partial \rho_{in}}{\partial v}\right|_{v*} - \left.\frac{\partial \rho_{st}}{\partial v}\right|_{v*}\right)\exp(-t/\tau)\right]$$
$$= \dot{N}_{\mathrm{st}} + (\dot{N}_{\mathrm{in}} - \dot{N}_{\mathrm{st}})\exp(-t/\tau) \tag{24}$$

where N˙inN˙in and N˙stN˙st are nucleation rates corresponding, respectively, to the initial and steady-state cluster distributions. The relaxation time of transient nucleation rate is identical with the relaxation time of cluster size distribution τ.

We consider amorphous melt superheated to the temperature $T_0 > T_m$ and rapidly quenched to a constant crystallization temperature $T < T_m$. Nucleation proceeds in the temperature Tstarting with an initial cluster size distribution in the melt equal to the equilibrium Boltzmann distribution ρin(v,T)=ρeq(v,T0)ρin(v,T)=ρeq(v,T0) at $T_0 > T_m$. The initial nucleation rate N˙inN˙in at T resulting from the equilibrium cluster distribution quenched from the overheating temperature T_0 reads

$$\dot{N}_{\mathrm{in}}(T) = -D[v^*(T), T]\left.\frac{\partial \rho_{eq}(v, T_0)}{\partial v}\right|_{v*(T)} \tag{25}$$

From Eq. (9), we have

$$\left.\frac{\partial \rho_{eq}(v, T_0)}{\partial v}\right|_{v^*(T)} = -\rho_0 \exp\left[-\frac{\Delta G(v^*(T), T_0)}{kT_0}\right]\frac{\Delta h(T_0 - T)}{kT_0 T_m} \tag{26}$$

while the steady-state nucleation rate N˙st(T)N˙st(T) is given by Eq. (18). The half-time t1/2(T)t1/2(T) at which isothermal transient nucleation rate reaches 50 % of the steady-state value reads

$$\frac{t_{1/2}}{\tau} = \ln\left[2\left(1 - \frac{\dot{N}_{\mathrm{in}}}{\dot{N}_{\mathrm{st}}}\right)\right] \tag{27}$$

Kashchiev [22] and Chakraverty [23] obtained the longest relaxation time in the form

$$\tau(T) = \frac{1}{2D[v^*(T)]\, Z^2(T)} \tag{28}$$

where D(v∗)D(v∗) and the Zeldovich factor Z(T)Z(T) are given by Eqs. (17) and (14), respectively. The relaxation time τ represents time of jumping of a critical-size cluster v^* over the thermodynamic barrier Z and is expressed by the following function of temperature

$$\tau(T) = \frac{192kT\sigma^3}{D(v*, T)\Delta g^4(T)} = \frac{4\sigma T_m^2}{\Delta h^2(T_m - T)^2 v_0^{4/3}} h \exp\left(\frac{E_D}{kT}\right) \tag{29}$$

ττ is controlled by thermodynamic driving force Δg(T)Δg(T) and the coefficient D(T)D(T).

Nucleation in non-isothermal conditions

Experimental observations and computer simulations indicate that time variation in thermodynamic parameters plays significant role in the nucleation kinetics. Continuity equation, Eq. (2), with time-dependent "diffusion coefficient" D and free enthalpy of cluster formation ΔG cannot be solved by separation of the variables. For a semiquantitative analysis of non-isothermal primary homogeneous nucleation, we consider an approximate single-relaxation-time model for time development of the cluster size distribution using temperature-dependent functions ρst(T)ρst(T) and τ(T)τ(T) in Eq. (23).

Transient non-isothermal nucleation rate

Consider temperature T monotonically varying in time in the range $T < T_m$ with the rate T˙=dT/dtT˙=dT/dt from T_{in}. Replacing time variable with temperature T(t)T(t) in Eq. (23), we obtain the following formula for variation in the transient cluster size distribution ρ(v,T)ρ(v,T) with temperature under non-isothermal conditions

$$\dot{T}(T)\frac{\partial\rho(v,T)}{\partial T} = \frac{\rho_{st}(v,T) - \rho(v,T)}{\tau(T)} \tag{30}$$

controlled by the relaxation time τ(T)τ(T) and steady-state limit ρst(v,T)ρst(v,T) of the distribution. For non-isothermal processes, the initial cluster size distribution at the start, $\rho_{in}(v)$, is assumed as steady-state distribution at the initial temperature T_{in} which is below T_m

$$\begin{aligned} t &= 0 \\ T &= T_{\text{in}} < T_m \\ \rho &= \rho_{\text{in}}(v) = \rho_{\text{st}}(v, T_{\text{in}}) \end{aligned} \tag{31}$$

By integration of Eq. (30) with time (temperature-)-dependent material functions and boundary conditions from Eq. (31), we obtain the following formula for temperature-dependent cluster size distribution with transient effects under cooling or heating

$$\begin{aligned} \rho_{\text{trans}}(v,T) &= \exp\left(-\int_{T_{\text{in}}}^{T} \frac{dT'}{\dot{T}(T')\tau(T')}\right) \\ &\times \left[\rho_{\text{in}}(v) + \int_{T_{\text{in}}}^{T} \exp\left(+\int_{T_{\text{in}}}^{T'} \frac{dT''}{\dot{T}(T'')\tau(T'')}\right) \frac{\rho_{st}(v,T')}{\dot{T}(T')\tau(T')} dT'\right] \end{aligned} \tag{32}$$

The transient effects in the cluster distribution under non-isothermal conditions, Eq. (32), are controlled by the product of the relaxation time and cooling/heating rate during the process, T˙(T)τ(T)T˙(T)τ(T), while under isothermal conditions discussed, they are controlled solely by τ(T)τ(T).

Asymptotic behavior of the transient cluster distribution in non-isothermal conditions can be characterized as follows

$$\rho_{\text{trans}}(v,T) = \left\{ \begin{array}{lll} \tau = 0; \dot{T} \neq 0 & \rho = \rho_{st}[v, T(t)] & \text{quasi - static} \\ \tau = 0; \dot{T} = 0 & \rho = \rho_{st}[v, T] & \text{isothermal, steady - state} \\ \tau \to \infty; \text{any}\dot{T} & \rho = \rho_{in}(v) & \text{quenched initial} \\ |\dot{T}| \to \infty; \text{any}\tau & \rho = \rho_{in}(v) & \text{quenched initial} \end{array} \right\} \tag{33}$$

Differentiation of the distribution terms in Eq. (32) with respect to cluster volume leads us to the following original formula for transient nucleation rate under non-isothermal conditions where the critical size v^* and the coefficient D(v∗)D(v∗) refer to instantaneous temperature T

$$\begin{aligned} \dot{N}_{\text{trans}}(T) &= -D(T,v*)\left[\left.\frac{\partial \rho_{\text{in}}(v)}{\partial v}\right|_{v*(T)} \exp\left(-\int_{T_{\text{in}}}^{T} \frac{dT'}{\dot{T}(T')\tau(T')}\right)\right. \\ &\left. + \int_{T_{\text{in}}}^{T} \left.\frac{\partial \rho_{\text{st}}(v,T')}{\partial v}\right|_{v*(T)} \exp\left(-\int_{T'}^{T} \frac{dT''}{\dot{T}(T'')\tau(T'')}\right) \frac{dT'}{\dot{T}(T')\tau(T')}\right] \end{aligned} \tag{34}$$

The initial distribution at the start, $\rho_{in}(v)$, in Eq. (34) is a steady-state distribution given by Eq. (13) at $Tin<TmTin<Tm$.

Consider the process in which cooling (heating) rate T˙T˙ is a constant and the relaxation time τvaries with temperature according to Eq. (29). Then, the instantaneous nucleation rate at Treduces to

$$\dot{N}_{\text{trans}}(T,\dot{T}) = -D(T,v*) \times \left[\left.\frac{\partial \rho_{\text{in}}(v)}{\partial v}\right|_{v*(T)} \exp\left(-\frac{1}{\dot{T}} \int_{T_{\text{in}}}^{T} \frac{\mathrm{d}T'}{\tau(T')} \right) + \frac{1}{\dot{T}} \int_{T_{\text{in}}}^{T} \left.\frac{\partial \rho_{st}(v,T')}{\partial v}\right|_{v*(T)} \exp\left(-\frac{1}{\dot{T}} \int_{T'}^{T} \frac{\mathrm{d}T''}{\tau(T'')} \right) \frac{\mathrm{d}T'}{\tau(T')} \right] \tag{35}$$

Nucleation rate in transient conditions is a function of three factors: undercooling Tm−TTm−T, temperature-dependent relaxation time, τ(T)τ(T), and cooling/heating rate, T˙T˙. The relaxation timeτ and cooling rate T˙T˙ affect nucleation kinetics in a similar way. The longer is τ and/or the faster is cooling, the closer is nucleation rate to the initial value. On the other hand, slow cooling rates and/or short relaxation times move nucleation rate in the direction of asymptotic steady state.

Thermal versus athermal nucleation

Besides *thermal* nucleation discusses above, under non-isothermal conditions there appears also another mechanism, *athermal nucleation*. The total concentration of supercritical clusters in a system is related to the cluster size distribution ρ(v,t)ρ(v,t)

$$N^*(t) = \int_{v*(t)}^{\infty} \rho(v,t)\mathrm{d}v \tag{36}$$

The rate with which N^* is changed, i.e., nucleation rate, results from the differentiation of the integral in Eq. (36) with respect to time

$$\dot{N}(t) = \frac{\mathrm{d}}{\mathrm{d}t}\int\limits_{v*(t)}^{\infty}\rho(v,t)\mathrm{d}v = \int\limits_{v*(t)}^{\infty}\frac{\partial\rho(v,t)}{\partial t}\mathrm{d}v - \frac{\mathrm{d}v^*}{\mathrm{d}t}\rho(v^*,t)$$

$$\int\limits_{v*(t)}^{\infty}\frac{\partial\rho(v,t)}{\partial t}\mathrm{d}v = j(v^*) - j(\infty) = j(v^*) = \dot{N}_{\mathrm{th}}(t) \tag{37}$$

$$\dot{N}_{\mathrm{ath}}(t) = -\frac{\mathrm{d}v^*}{\mathrm{d}t}\rho(v^*,t)$$

Differentiation of the kernel followed by the application of the Gauss–Ostrogradsky theorem yields flux of clusters leaving the subcritical region, $v < v^*$, and entering the region of stable nuclei, $v > v^*$. This mechanism is known as thermal nucleation. In steady-state isothermal conditions, the flux term j(v∗)j(v∗) reduces to N˙st(T)N˙st(T), Eq. (18). In variable external conditions, when critical cluster size v^* changes in time, there appears another term in Eq. (37) resulting from the differentiation of the lower integration limit, v∗(t)v∗(t). This term describes athermal nucleation which is proportional to the rate of external conditions (temperature, T, pressure, p, etc.)

$$\dot{N}_{\mathrm{ath}}(t) = -\frac{\mathrm{d}v^*}{\mathrm{d}t}\rho(v^*,t) = -\left[\frac{\partial v*}{\partial T}\dot{T} + \frac{\partial v*}{\partial p}\dot{p} + \cdots\right]\rho(v^*,t) \tag{38}$$

Depending on the sign of the derivative dv^*/dt, the athermal process may increase or reduce the number of nuclei in the system, e.g., cooling (T˙<0T˙<0) and/or compression (p˙>0p˙>0) virtually produces new nuclei, while heating or decompression associated with an increase in v^* makes some nuclei lose their stability. In the latter conditions, the athermal nucleation rate is negative. It should be mentioned that finite critical cluster size is defined only in the conditions of undercooling. Above the equilibrium transition temperature, $T > T_m$, the critical cluster size is infinite and no nucleation, thermal or athermal, is possible

$$\dot{N}_{\mathrm{ath}}(T) = \begin{cases} >0 & T<T_m;\dot{T}<0 \\ <0 & T<T_m;\dot{T}>0 \\ =0 & T=T_m \\ =0 & T>T_m \end{cases} \tag{39}$$

The concept of athermal nucleation has originally been introduced by Fisher, Hollomon, and Turnbull [40] and later derived in the form of Eq. (37) by Ziabicki [41] and Kashchiev [22]. Athermal nucleation is not controlled by diffusion and does not require thermal activation. Clusters acquire the status of nuclei under the change in external conditions. Reduction in temperature [46], or increase in pressure [47] shifts the criterion of cluster stability to smaller critical volume v^*, thus yielding virtual production of stable nuclei. In other words, some clusters, unstable in the conditions (T_1, p_1), are promoted to the rank of nuclei by changing thermodynamic state of the system to (T_2, p_2)

$$\begin{aligned} &T_1; v < v^*(T_1) \xrightarrow{\text{cooling}} T_2 < T_1; v > v^*(T_2) \\ &p_1; v < v^*(p_1) \xrightarrow{\text{compression}} p_2 > p_1; v > v^*(p_2) \end{aligned} \tag{40}$$

In the conditions of cooling and steady-state cluster distribution, ρst(v)ρst(v), we obtain the following formula for athermal nucleation rate from Eqs. (15) and (37)

$$\begin{aligned} &\dot{N}_{\text{ath}}(T, \dot{T}) = -\dot{T}\rho_{\text{st}}(v^*)\frac{dv^*}{dT} = -\dot{T}\Phi(T) \\ &\Phi(T) = \frac{96\sigma^3 T_m^3}{v_0^2 \Delta h^3 (T_m - T)^4} \exp\left[\frac{-32\sigma^3 T_m^2}{kT\Delta h^2 (T_m - T)^2}\right] \end{aligned} \tag{41}$$

Replacing ρst(v)ρst(v) in Eq. (41) by transient cluster size distribution ρtrans(v,T)ρtrans(v,T) given by Eq. (32), an analytical closed-form expression for athermal nucleation rate is obtained, accounting for the relaxation effects.

It should be emphasized that both thermal and athermal nuclei are homogeneous by nature and physically undistinguishable. Equation (41) accounts for athermal effects in sporadic homogeneous nucleation, without predetermined homogeneous nucleation. To account for the effects of predetermined nucleation, for example memory effects, the distribution ρst(v∗)ρst(v∗) in Eq. (41) should be modified for the predetermined clusters contribution, what is not a subject of this paper.

Concentration of nuclei produced in a single cooling/heating run

The factor subsequent to nucleation which controls crystallization is the concentration of nuclei produced during cooling or heating. For steady-state isothermal processes, concentration of thermal nuclei produced within the time interval Δt is

$$\Delta N_{\mathrm{st}}(\Delta t) = \dot{N}_{\mathrm{st}}(T) \cdot \Delta t \tag{42}$$

Consider a single cooling/heating run starting at T0<TmT0<Tm. With transient effects in non-isothermal process, we obtain the following closed-form analytical formula for the concentration of nuclei produced in the temperature interval (T0T0, *T*)

$$\Delta N(T) = \int_{T_0}^{T} \frac{\dot{N}_{\mathrm{trans}}(T', \dot{T})\mathrm{d}T'}{\dot{T}(T')} \tag{43}$$

where N˙trans(T′,T˙)N˙trans(T′,T˙) is given by Eq. (35).

In the nucleation considered as a quasi-static process (steady-state nucleation rate instantaneously follows varying temperature), concentration of nuclei produced at constant T˙T˙, at the absence of transient and athermal effects, is expressed by the integral

$$\Delta N_{\mathrm{qs}}(T) = \frac{1}{\dot{T}} \int_{T_0}^{T} \dot{N}_{\mathrm{st}}(T')\mathrm{d}T' \tag{44}$$

N˙st(T)N˙st(T) is a positive material function, and the concentration of nuclei produced in a quasi-static run is controlled by the temperature interval and the cooling/heating rate. For both cooling and heating, concentration of nuclei produced in the transient or quasi-static processes increases, Eqs. (43) and (44). Cooling or heating below T_m provides positive contribution to the concentration of thermal nuclei.

Different behavior is found for athermal nucleation. Integrating the athermal nucleation rate in the quasi-static limit, Eq. (41), concentration of athermal nuclei produced in the run is obtained, independently of the cooling/heating rate T˙T˙

$$\Delta N_{\mathrm{ath}}(T) = -\int_{T_0}^{T} \Phi(T')\mathrm{d}T' \tag{45}$$

Unlike for thermal nucleation, contribution of athermal process is different for cooling and heating. Under cooling, athermal nuclei are produced, ΔNath(T)>0ΔNath(T)>0, while heating results in reduction in the nuclei concentration, ΔNath(T)<0ΔNath(T)<0.

Total concentration of nuclei at *T* created in the system during a single cooling/heating run in quasi-static process is expressed by the sum

$$\begin{aligned}\Delta N(T) &= \Delta N_{\mathrm{qs}}(T) + \Delta N_{\mathrm{ath}}(T)\\ &= \int_{T_0}^{T} \frac{\dot{N}_{\mathrm{st}}(T')\mathrm{d}T'}{\dot{T}} - \int_{T_0}^{T} \Phi(T')\mathrm{d}T'\end{aligned} \tag{46}$$

It is evident that the thermal contribution, ΔNqsΔNqs, is inversely proportional to the cooling/heating rate, while the athermal contribution depends only on the temperature interval (T_0, T).

Discussion of the results

The model in single-relaxation-time approximation allows to predict transient nucleation rate under isothermal and non-isothermal conditions, athermal nucleation kinetics, as well as influence of individual material and processing parameters and concentration of nuclei produced in a cooling or heating run. Concentration of nuclei produced in a single cooling/heating run is controlled by initial and target temperature and the cooling/heating rate. The effect of temperature is associated with the driving force represented by the product of heat of melting and the degree of undercooling. Cooling rate renormalizes timescale into temperature (Eq. 43) and affects transient thermal nucleation rate through the product with the relaxation time (Eq. 34). Transient effects present in isothermal and non-isothermal processes are associated with activation energy via the relaxation time.

Variation in cluster size distribution provides a basis for nucleation rate and is assumed to be controlled by the longest relaxation time. Thermodynamic driving force is considered as a linear function of undercooling; molecular transport effects are described by the

Arrhenius formula with a constant activation energy E_D. The range of temperatures available for thermal nucleation is limited to Tlim≤T<TmTlim≤T<Tm for PHB. When temperature T_{lim} is reached, all clusters including monomers become effective nuclei and the kinetics of crystallization is solely controlled by diffusional growth. For indium, the nucleation mechanism is active in the entire range of temperatures due to much smaller molecular volume of the kinetic units.

Besides thermal nucleation mechanism, there appears athermal nucleation. The thermal and athermal nuclei are homogeneous by nature and physically undistinguishable. Athermal nuclei do no require thermal activation of molecular transport because they are created due to a change in the critical cluster size under changing temperature.

Isothermal nucleation rate in this model with a single-relaxation time consists of a steady-state and transient terms. Figure 1 shows steady-state nucleation rate N˙stN˙st versus temperature *T*calculated from Eq. (18) for indium and Fig. 2 the same characteristics for PHB. Thermodynamic barrier to nucleation, ΔG^*, approaches infinity at $T = T_m$ and determines upper limit of nucleation. With increased undercooling, ΔG^* is reduced and the thermodynamic driving force increases. At the same time, cooling suppresses molecular mobility and slows nucleation down. Competition between the thermodynamic and transport effects leads to a maximum of N˙st(T)N˙st(T) at $T = T_{\text{max}}$. The calculated temperature T_{max} for PHB is 288 K, i.e., lies beyond the range of nucleation, Eq. (7). Virtual continuation of the function N˙st(T)N˙st(T) below $T_{\text{lim}}$is presented as a dashed curve in Fig. 2. For indium, $T_{\text{max}} = 162$ K and lies within the range of nucleation.

Nucleation rates, very small in the vicinity of melting temperature (Figs. 1, 2), increase with increasing undercooling. N˙st(T)N˙st(T) for the investigated materials covers wide range of nucleation rates from zero to 10^{22}–10^{26} cm^{-3}s^{-1}. The pre-exponential factor I_0 calculated from Eq. (18) is a weak function of temperature. For indium, the calculated value of I_0 slightly increases from 2.37×10^{35} cm^{-3}s^{-1} at 280 K to 2.83×10^{35} cm^{-3}s^{-1} at 400 K. This is well consistent with the average value $I_0 \approx 10^{35\pm1}$ cm^{-3}s^{-1} reported from experimental investigations on liquid metals [19]. For PHB, the calculated factor I_0 amounts to 0.87×10^{34} cm^{-3}s^{-1} at 350 K and 1.11×10^{34} cm^{-3}s^{-1} at 440 K. Experimental data on nucleation for polymers analyzed by Hoffman [3] yield $I_0$of the same order, i.e., 10^{34} cm^{-3}s^{-1}.

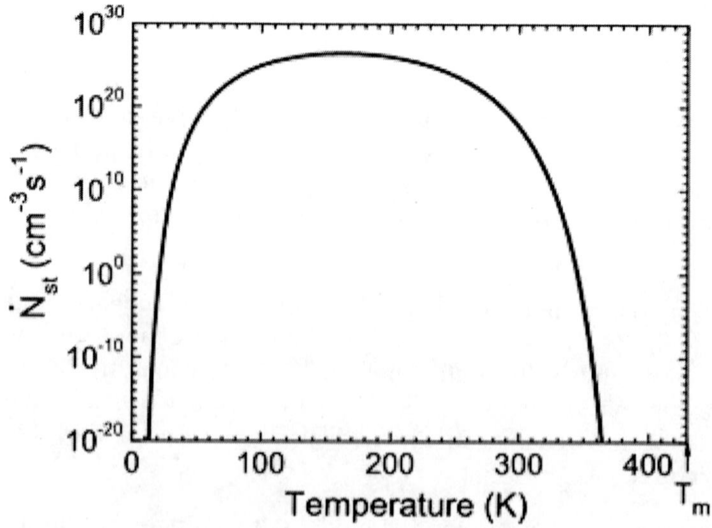

Figure 1. Steady-state nucleation rate N˙stN˙st for indium versus temperature *T*. TmTm—equilibrium melting temperature

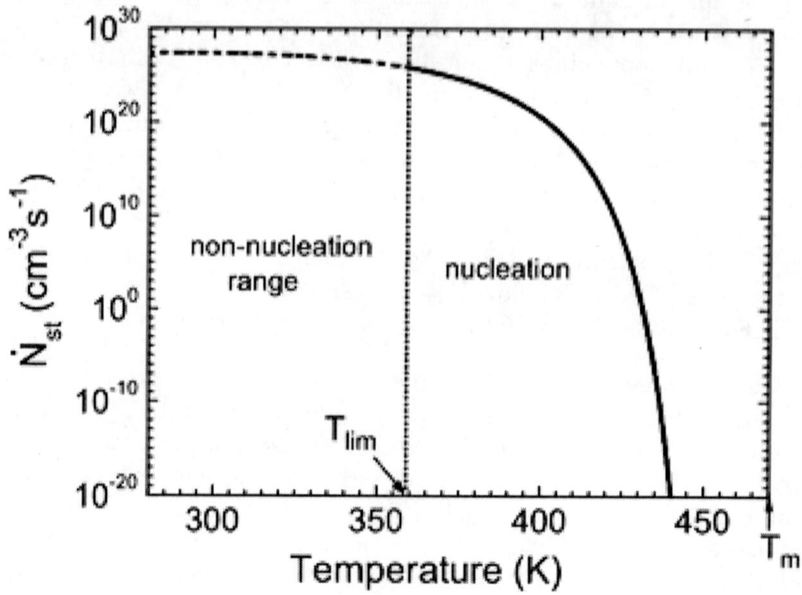

Figure 2. Steady-state nucleation rate N˙stN˙st for PHB versus temperature *T* in the range well above glass transition. TmTm—equilibrium melting temperature, TlimTlim—limiting temperature for nucleation. The plot in the non-nucleation range indicated by dashed line

Figures 3 and 4 show transient isothermal nucleation rates reduced by their steady-state values, N˙trans(t)/N˙stN˙trans(t)/N˙st, calculated from Eqs. (24) and (18) versus time *t* for indium and PHB. The plots are shown for different temperatures *T* and corresponding nucleation relaxation times *τ*, calculated with the initial equilibrium Boltzmann distribution of clusters quenched to *T* from T0=Tm+10KT0=Tm+10K. The isothermal transient nucleation rate starts from a very low level, N˙in≅0N˙in≅0, determined by the quenched initial cluster distribution from T0T0 and increases in time asymptotically approaching steady-state level N˙stN˙st. The plots for indium and PHB are shown for three nucleation temperatures and corresponding nucleation relaxation times. The higher is nucleation temperature and longer is the relaxation time, the more nucleation rate does lag behind the steady-state value.

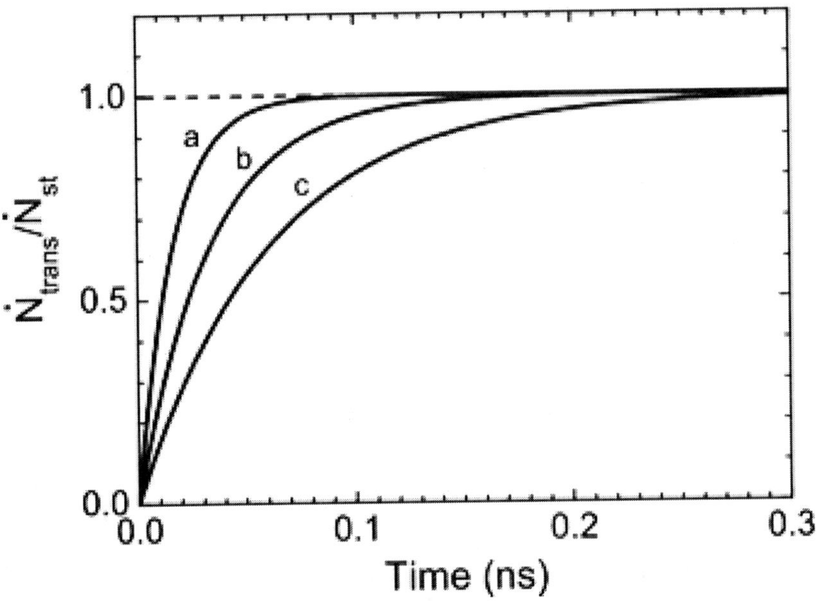

Figure 3. Reduced isothermal transient nucleation rates N˙trans/N˙stN˙trans/N˙st versus time *t* for indium. **a** *T* = 320 K,*τ* = 0.016 ns; **b***T* = 360 K, *τ* = 0.033 ns; **c***T* = 380 K, *τ* = 0.061 ns. Rapid quench to *T* from T0=Tm+10KT0=Tm+10K

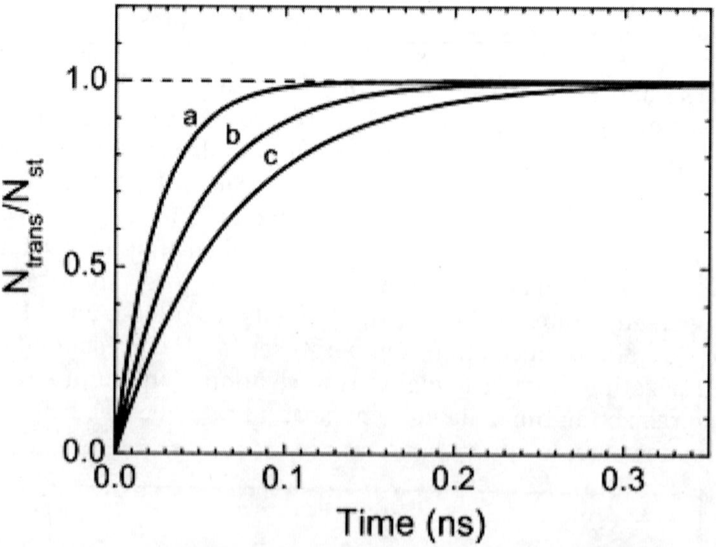

Figure 4. Reduced isothermal transient nucleation rate N˙trans/N˙stN˙trans/N˙st versus time *t* for PHB. **a** T = 400 K,τ = 0.025 ns; **b** T = 430 K, τ = 0.045 ns; **c** T = 440 K, τ = 0.069 ns. Rapid quench to T from T0=Tm+10KT0=Tm+10K

Figure 5 illustrates temperature dependence of the reduced halftime of isothermal transient nucleation rate, t1/2/τt1/2/τ, calculated from Eq. (27) for indium and PHB. For each material, the plots for various overheating temperatures T_0 overlap. In the range of high temperatures (above 200 K for indium and above T_{lim} for PHB), the reduced halftime is nearly ln2 which means that in this temperature range, contribution of the initial cluster distribution quenched from T_0above TmTm is negligible. In lower temperatures, the halftime for indium rapidly drops to zero because the initial nucleation rate resulting from the quenched distribution tends to one half of the steady-state value, Eq. (27). For PHB, variation in the reduced halftime in the low-temperature range is beyond the nucleation mode.

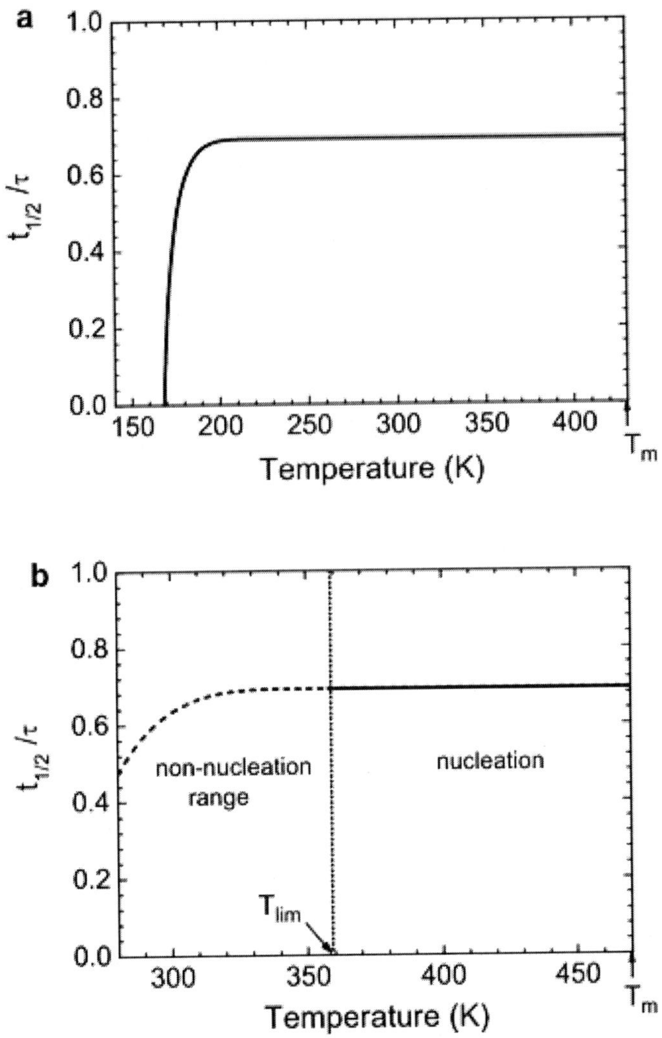

Figure 5. Reduced halftime of the transient nucleation rate t1/2/τt1/2/τ versus *T* calculated for—**a** indium, **b** PHB. Rapid quench from the overheating T0−Tm=10,30,50KT0−Tm=10,30,50K(the plots overlap). *Dashed plot*—the non-nucleation range

Under non-isothermal conditions, thermal nucleation rate is influenced by the product of cooling rate and temperature-dependent relaxation time, T˙τT˙τ. Nucleation rate is reduced by the relaxation effects the more, the longer is relaxation time and the faster is cooling. Under slow cooling and/or short relaxation times, nucleation rate approaches asymptotically steady state. Temperature dependence of the relaxation time τ and transient nucleation rates for different values of the product T˙τT˙τ computed for indium and PHB from Eqs. (29) and (35) are presented in Fig. 6. In the melting temperature T=TmT=Tm, Δg is zero and the relaxation time is infinitely long. The function τ(T)τ(T) exhibits a minimum and increases in the range of large undercoolings. The shortest relaxation time calculated for indium, τ = 0.009 ns, is found to appear at 230 K and that for PHB, τ = 0.020 ns, at 374 K.

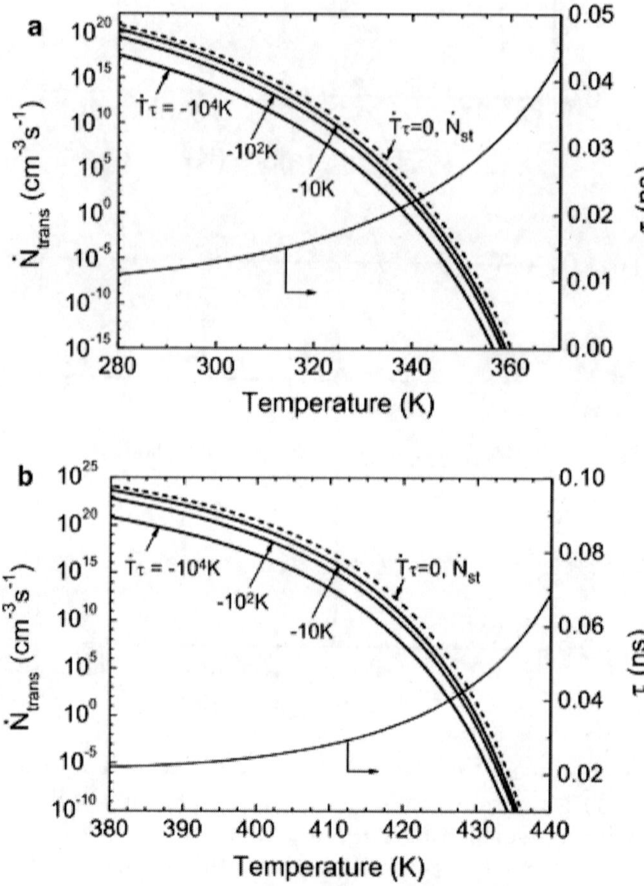

Figure 6. Non-isothermal transient nucleation rates N˙transN˙trans and relaxation time τ versus temperature T for—**a**indium, **b** PHB. Parameter T˙τT˙τ indicated. Initial temperature of non-isothermal nucleation Tin=Tm−0.01KTin=Tm−0.01K. *Dashed line*—nucleation rate in quasi-static conditions, T˙τ=0T˙τ=0

Scanty literature data on the relaxation time in primary crystal nucleation cover wide range of values [5] including subnanosecond values calculated from Eq. (29) for liquid indium and molten PHB. Processes controlled by so short relaxation times seem to have no effect on crystal nucleation in liquid-to-solid transitions. On the other hand, crystallization from the solid or glassy state involves relaxation times as long as 10^2–10^5 s [5, 31, 48], and in such systems, relaxation does play critical role in the nucleation kinetics.

Transient nucleation rates $\dot{N}trans(T)\dot{N}trans(T)$ are the smaller, the larger is the product $||\dot{T}\tau|||\dot{T}\tau|$. Taking into account that relaxation times in the liquid–solid transitions are in the nanosecond range (Fig. 6) and practically attainable cooling/heating rates do not exceed 10^6 K/s, the product $||\dot{T}\tau|||\dot{T}\tau|$ is smaller than 10^{-3} K. Consequently, effect of relaxation is weak and in most cases can be neglected. When the relaxation time τ is equal to zero and cooling rate $\dot{T}\dot{T}$ is different from zero, the nucleation process is quasi-static. Then, cluster distribution and nucleation rate follow variation in temperature and in every instant of time assume the steady-state value (dashed lines in Fig. 6)

$$\begin{aligned} \rho_{\mathrm{trans}}[v, T(t)] &\xrightarrow{\tau \to 0} \rho_{\mathrm{st}}[v, T(t)] \\ \dot{N}_{\mathrm{trans}}[T(t)] &\xrightarrow{\tau \to 0} \dot{N}_{\mathrm{st}}[T(t)] \end{aligned} \tag{47}$$

Kelton [7, 34] discussed homogeneous crystal nucleation in a metallic glass $Au_{81}Si_{19}$ as a function of temperature and cooling rate. Maximum nucleation rate of the order of 10^{18} $cm^{-3}s^{-1}$in the steady state $(\dot{T}=0)(\dot{T}=0)$ is reduced to ca 10^4 $cm^{-3}s^{-1}$ at $\dot{T}=-107K/s\dot{T}=-107K/s$. Similar behavior is evident in Fig. 6 where the nucleation rates for indium and PHB plotted with $\dot{T}\tau=const\dot{T}\tau=const$ are reduced by three orders of magnitude when steady-state conditions $(\dot{T}\tau=0\dot{T}\tau=0)$ are replaced with $\dot{T}\tau=-104K\dot{T}\tau=-104K$.

Figure 7 presents athermal nucleation rates for indium and PHB computed from Eq. (41) as functions of temperature T at constant cooling rates. It is evident that athermal nucleation rate is a function of temperature and cooling rate. Dashed line in each figure presents steady-state thermal nucleation rate, Eq. (18). $\dot{N}ath\dot{N}ath$ increases with increasing undercooling, $Tm-TTm-T$, and with increasing cooling rate. Unlike transient nucleation rates (Fig. 6), steady-state thermal nucleation rate does not provide any bound for athermal nucleation.

The function N˙ath(T)N˙ath(T)starts from zero at T˙=0T˙=0 and approaches infinity at infinite cooling rate.

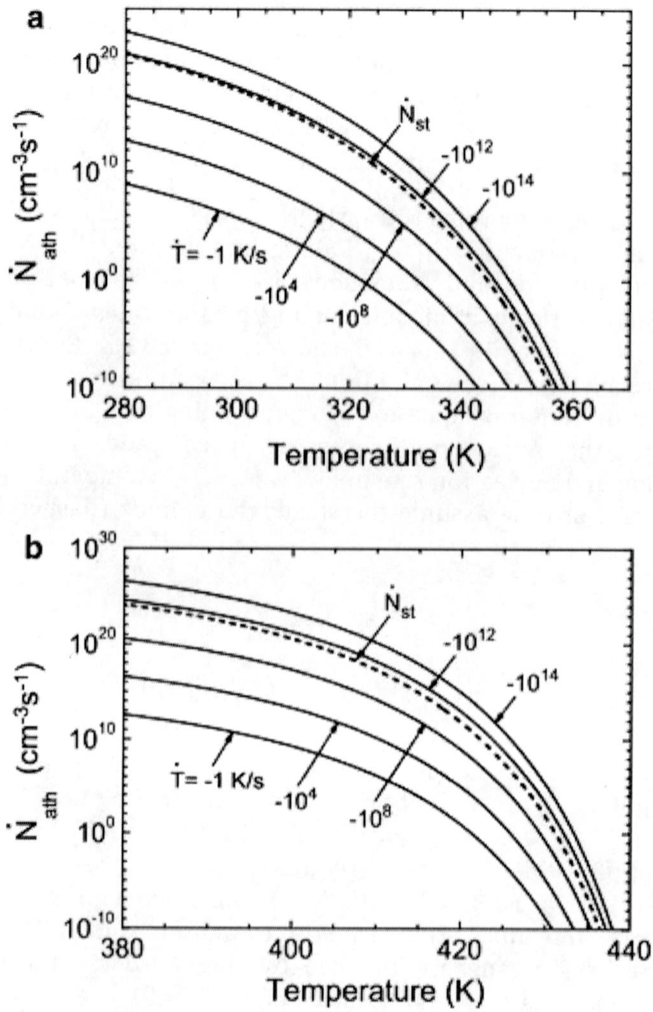

Figure 7. Athermal nucleation rates N˙athN˙ath versus temperature *T* for—**a** indium, **b** PHB. Cooling rates T˙T˙indicated. *Dashed line*—steady-state thermal nucleation rate N˙stN˙st

Steady-state thermal nucleation rate provides boundary for transient nucleation mechanism and not for athermal nucleation. At some cooling rates, T˙=T˙cross(T)T˙=T˙cross(T), athermal nucleation rate crosses steady-state thermal nucleation rate, and both mechanisms have the same influence on nucleation. For athermal nucleation under quasi-static conditions, we obtain

$$\dot{N}_{\text{ath}}[T, \dot{T}_{\text{cross}}(T)] = \dot{N}_{\text{st}}(T)$$
$$\dot{T}_{\text{cross}}(T) = \frac{-\Delta h(T_m - T)^2 \sqrt{kT}}{4\sqrt{6\pi}\sigma^{3/2} T_m} \frac{1}{\tau} \quad (48)$$

The athermal nucleation overruns the thermal process at the cooling rates exceeding T˙crossT˙crosswhich is inversely proportional to the relaxation time τ and increases with increasing the undercooling, Tm−TTm−T. The function N˙ath(T)N˙ath(T) starts from zero at T˙=0T˙=0 and approaches infinity at infinite cooling rate. With decreasing temperature (T˙<0T˙<0), the crossover cooling rate and athermal nucleation rate at the crossover, N˙ath[T,T˙cross(T)]N˙ath[T,T˙cross(T)], predicted from Eq. (48) increase. For example, in the case of indium at T=337KT=337K, we have N˙ath(T)=N˙st(T)=105cm−3s−1N˙ath(T)=N˙st(T)=105cm−3s−1at T˙cross=−1.8×1011 K/sT˙cross=−1.8×1011 K/s. The same nucleation rates for PHB are achieved at T=427KT=427Kand T˙cross=−4.4×1010 K/sT˙cross=−4.4×1010 K/s. At small undercoolings, the crossover nucleation rates are negligibly small in both cases. At deeper undercoolings, the cooling rates required for the crossover are extremely high due to high thermal nucleation rates exceeding experimentally available range. This behavior predicted for the molten metal (indium) and for the polymer melt (PHB) seems to be typical for liquid-to-solid transitions characterized by very short relaxation times.

Contribution of athermal nucleation (positive or negative) in case of pure sporadic nucleation is negligibly small for experimentally controlled cooling rates, compared to thermal mechanism. Significant contribution of athermal nucleation might be expected in processes with predetermined homogenous nuclei, for example memory effects, which modify the cluster size distribution.

Total concentration of nuclei created in a single cooling/heating run is the sum of integrals of thermal nucleation rate reduced by cooling/heating rate (Eq. 43), and athermal contribution controlled by temperature interval (T0−T)(T0−T) but independent of the cooling/heating rate (Eq. 45). Thermal nucleation mechanism provides positive contribution to the concentration of nuclei both for cooling and heating. Athermal nucleation, on the other side, yields positive contribution at cooling and negative at heating.

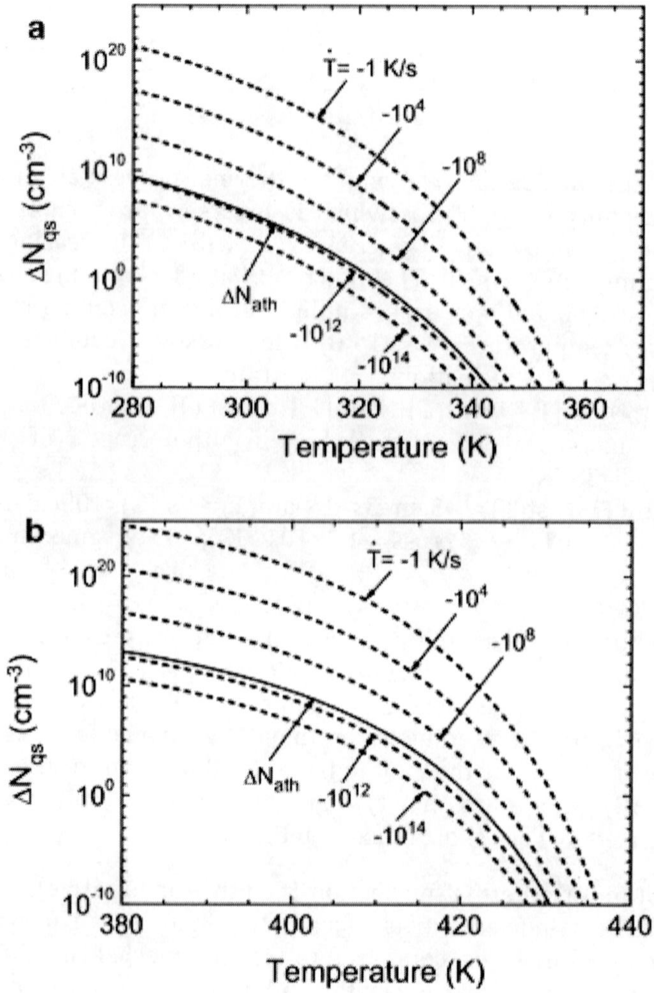

Figure 8. Concentration of nuclei produced in quasi-static process ΔNqsΔNqs versus temperature *T* (*dashed lines*) at various cooling rates (indicated) for—**a** indium, **b** PHB. *Solid line*—gain in the nuclei concentration due to athermal nucleation. T0=Tm−0.01KT0=Tm−0.01K

Figure 8 illustrates the concentration of nuclei ΔNqs(T)ΔNqs(T) produced in a quasi-static process during cooling at various cooling rates for indium and PHB, calculated from

Eq. (46) at the initial temperature T0= Tm−0.01KT0= Tm−0.01K (dashed lines). Solid lines in Fig. 8 show a gain in the concentration of nuclei due to the athermal quasi-static process, ΔNath(T)ΔNath(T). At higher cooling rates, lower increase in the concentration of nuclei produced in thermal mechanism is obtained at any temperature *T*, while concentration of athermal nuclei created is independent of the cooling rate.

Effects of material properties

Two model materials: molten elemental indium and molten organic polymer (PHB) are discussed. The material characteristics are collected in Table 1. In spite of different molecular structures, kinetics of nucleation behavior is qualitatively similar for both materials. Basic differences between molten indium and PHB involve the size of single kinetic units, density, heat of melting, equilibrium melting temperature, molecular mobility, and interface free energy density.

For indium, the kinetic units considered are single atoms, and for PHB statistical Kuhn segments with the volume ca 100 times larger than indium atoms. Although heat of fusion of indium is 13 % greater than that of PHB, its melting temperature is lower by about 40 K than that of PHB. Melting entropy of the polymeric material is smaller because of some intramolecular order in the molten state. Molecular mobility of PHB approximated by the Arrhenius formula is much smaller than that of indium due to the activation energy E_D nearly six times higher for the polymer. Interface free energy densities, nearly twice higher for indium, affect thermodynamic barrier of nucleation, ΔG^*.

Nucleation rates start as zero at the respective melting temperatures T_m and increase with increasing undercooling, ΔT= Tm−TΔT=Tm−T. To obtain realistic homogeneous nucleation rates (say, 1 $cm^{-3}s^{-1}$), indium should be undercooled by about 80 K below T_m, while PHB requires twice lower undercooling, about 40 K. The highest nucleation rates calculated for the considered materials approach 10^{22}–10^{26} $cm^{-3}s^{-1}$.

Free energy barrier of primary homogeneous nucleation ΔG∗ΔG∗ at undercooling ΔTΔT is influenced by the interface free energy density σ and the ratio Δh/TmΔh/Tm which is the specific entropy of fusion, ΔsΔs. Higher value of σ for indium (Table 1) leads at any undercooling ΔTΔT to higher free energy barrier ΔG∗InΔGIn∗ despite higher Δh/TmΔh/Tm. At the undercooling ΔTΔT, the ratio of free energy barriers of indium and PHB

$$\frac{\Delta G^*_{\mathrm{In}}(\Delta T)}{\Delta G^*_{\mathrm{PHB}}(\Delta T)} = \left(\frac{\Delta s_{\mathrm{PHB}}}{\Delta s_{\mathrm{In}}}\right)^2 \left(\frac{\sigma_{\mathrm{In}}}{\sigma_{\mathrm{PHB}}}\right)^3 \qquad (49)$$

For the parameters listed in Table 1, we have ΔsPHB/ΔsIn=0.809ΔsPHB/ΔsIn=0.809, σIn/σPHB=1.872σIn/σPH B=1.872,
and ΔG∗In(ΔT)/ΔG∗PHB(ΔT)=4.29ΔGIn∗(ΔT)/ΔGPHB∗(ΔT)=4.29 at any undercooling. The transport activation energy is considerably lower for indium, EInD/EPHBD=0.180EDIn/EDPHB=0.180, and reduces the nucleation kinetics much less than for the polymer. Considerably lower volume of single elements for indium, vIn0/vPHB0=1.115×10−2v0In/v0PHB=1.115×10−2, leads to much higher pre-exponential factor I0I0 in Eq. (18), enhanced by higher σInσIn in the power of 1/2.

Analysis of individual influence of σ, Δh/TmΔh/Tm, EDED, and v0v0 indicates that despite much higher pre-exponential and transport terms for indium, the steady-state nucleation rate, Eq. (18), is lower than that for PHB by many orders of magnitude due to higher value of σ for indium. Higher σ results in steady-state nucleation rate for indium lower by 22 orders of magnitude at high undercooling (ΔT=100KΔT=100K). With decreasing undercooling, the reduction in the nucleation rate increases, and for ΔT=35KΔT=35K, it reduces by 151 orders of magnitude. The effect of higher values of Δh/TmΔh/Tm for indium is the steady-state nucleation rate higher by nine orders of magnitude at ΔT=100KΔT=100K which increases with decreasing undercooling to 70 orders of magnitude at ΔT=35KΔT=35K. This indicates that higher Δh/TmΔh/Tm for indium does not compensate much higher reduction in thermal nucleation kinetics caused by higher σ. Lower values of EDEDand v_0 for indium only slightly rise steady-state nucleation rate by three orders of magnitude and by a factor of 20, respectively, in the discussed range of undercooling.

The ratio of the relaxation time of indium and PHB, important for the transient effects (Eq. 29), at the undercooling ΔTΔT reads

$$\frac{\tau_{\mathrm{In}}(\Delta T)}{\tau_{\mathrm{PHB}}(\Delta T)} = \frac{\sigma_{\mathrm{In}}}{\sigma_{\mathrm{PHB}}}\left(\frac{\Delta s_{\mathrm{PHB}}}{\Delta s_{\mathrm{In}}}\right)^2\left(\frac{v_0^{\mathrm{PHB}}}{v_0^{\mathrm{In}}}\right)^{4/3} \times \exp\left[\frac{E_D^{\mathrm{In}}}{k(T_m^{\mathrm{In}} - \Delta T)} - \frac{E_D^{\mathrm{PHB}}}{k(T_m^{\mathrm{PHB}} - \Delta T)}\right] \tag{50}$$

The value of EDED lower for indium by about six times makes the relaxation time by three orders of magnitude lower than for the polymer. The interface free energy σ contributes to τ linearly and does not have such a dominant influence as it has on steady-state nucleation rate. Much smaller volume v_0 for indium results in high value of 498 of the pre-exponential factor in Eq. (50). Minor influence of σ and Δh/TmΔh/Tm on the difference in the relaxation time of indium and PHB is predicted. Lower EDED and v0v0 values for indium compensate each other in the influence on the difference in the relaxation time of both materials. The estimated ratio τIn(ΔT)/τPHB(ΔT)τIn(ΔT)/τPHB(ΔT) varies between slightly below unity at high undercooling and about 3 at low undercooling. This indicates similarity of the transient effects in the nucleation kinetics in a wide range of temperature for both materials.

The ratio of athermal nucleation rate of both materials at the undercooling ΔTΔT and temperature rate T˙T˙ is given by the ratio of functions $\dot{N}_{\mathrm{ath}}^{\mathrm{In}}/\dot{N}_{\mathrm{ath}}^{\mathrm{PHB}} = \Phi_{\mathrm{In}}(T_m - \Delta T)/\Phi_{\mathrm{PHB}}(T_m - \Delta T)$, and is influenced by σ, Δh/TmΔh/Tm and v0v0. The parameters influence the free energy barrier in Φ(T)Φ(T) higher for indium mainly due to higher value of σ and reduce the athermal nucleation rate ratio by many orders of magnitude. Much lower value of v0v0 for indium results in a high value of the ratio of the pre-exponential factors $\left(\frac{v_0^{\mathrm{PHB}}}{v_0^{\mathrm{In}}}\right)^2\left(\frac{\sigma_{\mathrm{In}}}{\sigma_{\mathrm{PHB}}}\right)^3\left(\frac{\Delta s_{\mathrm{PHB}}}{\Delta s_{\mathrm{In}}}\right)^3 = 2.80 \times 10^4$, which does not compensate strong influence of σ on the free energy barrier and the athermal nucleation kinetics. Higher σ for indium reduces athermal nucleation rate by 21 orders of magnitude at high undercooling (ΔT=100KΔT=100K), and the reduction increases with decreasing undercooling to 151 orders of magnitude at ΔT=35KΔT=35K. Contrary to the effect of higher value of σ, higher entropy of fusion Δh/TmΔh/Tm for indium only partly compensates the influence of σ and rises athermal nucleation rate from 9 to 69 orders of magnitude with decreasing undercooling from ΔT = 100–35 K. Similar influence of higher σ and Δh/TmΔh/Tm of indium is predicted also for steady-state nucleation kinetics.

The data for indium are shifted in the temperature scale by some 70 K with respect to PHB (Figs. 1, 2). Compared at constant temperature, or at constant undercooling, homogeneous nucleation rates are higher for PHB because of higher thermodynamic driving force of crystallization. At the same time, experimental observations for indium [49] show crystallization onset at the undercooling as low as 10–25 K. This effect has been identified as heterogeneous nucleation on the bulk impurities. According to the authors [49], much deeper undercooling is required for homogeneous nucleation. In a bacterial PHB free from impurities, Barham [50] observed homogeneous crystal nucleation under high undercoolings.

Summing up, kinetics of nucleation in molten indium and PHB are similar and seem to be representative of metals, polymers, and other materials undergoing liquid–solid transition. Different behavior has been observed in glass-to-solid and solid-to-solid transitions [7, 31].

Effects of processing conditions

Nucleation in the liquid–solid transitions is controlled by the thermodynamic variables (temperature, pressure, etc.) and variation rates of these variables (cooling/heating rate, compression rate, etc.). The thermodynamic potentials as functions of the variables determine phase equilibria and the range of conditions in which individual transitions take place. Undercooling determines thermodynamic driving force of the transition, $\Delta g(T)=-\Delta h(Tm-T)/Tm\Delta g(T)=-\Delta h(Tm-T)/Tm$, affects the rate of cluster formation, and enhances nucleation rates and the relaxation phenomena. The rates of temperature, pressure, and other functions of state play important role in the kinetics of nucleation. In this paper, we have concentrated on the effects of temperature and temperature rate in isobaric systems.

Nucleation rate is controlled by the competition of thermodynamic influence of undercooling and molecular mobility of the kinetic units subjected to clustering. The measure of molecular mobility is temperature-dependent diffusion coefficient in the space of cluster volumes, $D(v,T)D(v,T)$. In this paper concerned with nucleation in a liquid phase, the Arrhenius approximation was used, $D(v,T)\propto\exp(-ED/kT)D(v,T)\propto\exp(-ED/kT)$ with a constant transport barrier, E_D. For polymers subjected to deeper undercooling, close to glass transition T_g, the Arrhenius approximation may be replaced with Williams–Landel–Ferry formula

[51] $D(v,T) \propto \exp\left[\frac{C_1(T-T_g)}{C_2+T-T_g}\right]$ or Hoffman–Lauritzen formula [3], $D(v,T) \propto \exp\left[\frac{-E_D}{k(T-T_\infty)}\right]$, where $T_\infty = T_g - C_3$ –C3 and C1,C2,C3C1,C2,C3 are material constants.

The model indicates how the processing conditions can be adjusted to produce desired nucleation structure. Consider conditions required to minimize the concentration of nuclei to get nearly amorphous material. (1) The starting structure should be amorphous. This requires superheating at a temperature T0>TmT0>Tm long enough to erase structural memory [52, 53]. (2) The cooling rate T˙T˙ should be adjusted as high as possible. Complete elimination of the production of thermal nuclei is not possible as it would require infinite cooling rate. (3) The target temperature, *T*, should be adjusted according to the material involved. *T* must be higher than limiting level, T>TlimT>Tlim (Eq. 7). For a given material characterized by steady-state nucleation rate N˙st(T)N˙st(T) (cf. Figs. 1, 2), the target temperature should be chosen so that production of thermal nuclei is minimized (see Eq. 43). (4) In addition to thermal nucleation, cooling of a superheated melt provides concentration of athermal nuclei independent of cooling rate and increasing with the temperature interval T0–TT0–T (see Eq. 45). Ideal amorphous structure without any nuclei below TmTm cannot be obtained for two reasons: finite cooling rate and athermal nucleation.

Maximization of the concentration of homogeneous (thermal and athermal) nuclei can be performed according to a similar scheme. The starting temperature T0T0, cooling (heating) rate, and the initial structure are immaterial. One can produce nuclei by cooling superheated amorphous melt, or by heating partly nucleated material to the target temperature. The fastest way consists in cooling or heating the material to the temperature corresponding to the maximum nucleation rate (Figs. 1, 2). At the target temperature, further nucleation is performed isothermally. In the conditions when thermal and athermal nucleation does not provide expected concentration of nuclei, heterogeneous nuclei can be added to the system.

CONCLUSIONS

A kinetic model of homogeneous non-isothermal crystal nucleation with transient and athermal effects is developed. Transient effects are characterized by a single (the longest) relaxation time. Closed-form analytical formulas for transient cluster size distribution, related thermal and athermal nucleation rates, and the total number of nuclei produced in a cooling or heating run are derived. Under isothermal conditions, the transient term controlled by the relaxation time contributes to nucleation rate the more the higher is temperature. The longer is relaxation time and the smaller is undercooling, the more isothermal nucleation rate lags behind the steady state (Figs. 3, 4).

Under non-isothermal conditions, relaxation effects in thermal nucleation rate are controlled by the product of cooling (heating) rate and temperature-dependent relaxation time, $\dot{T}\tau\dot{T}\tau$. The cooling rate $\dot{T}\dot{T}$ and relaxation time τ affect nucleation kinetics in a similar way, the longer is $\tau$and/or the faster is cooling, the closer is nucleation rate to the initial value. Under slow cooling and/or short relaxation times, nucleation rate approaches asymptotically steady state (Fig. 6). Nucleation rate is reduced by relaxation effects the more, the longer is relaxation time and the faster is cooling.

Our model applied to molten indium and the linear polymer PHB, representing molten metals and polymers undergoing liquid-to-solid transitions, are characterized by relaxation times in the nanosecond range. This indicates that transient effects in nucleation would require extremely fast cooling of such materials. Material parameters strongly affecting kinetics of homogeneous nucleation include interface free energy density σ and the ratio $\Delta h/Tm\Delta h/Tm$. The highest impact on the nucleation rates results from considerably higher σ for indium which reduces steady-state and athermal nucleation rates by many orders of magnitude, relative to those predicted for the polymer. The reduction is stronger at lower undercooling. Higher value of $\Delta h/Tm\Delta h/Tm$ for indium results in a rise of the nucleation rates by some orders of magnitude, but it does not compensate the reduction caused by the σ value.

The predicted relaxation time τ is of the same order of magnitude for indium and PHB in a wide range of undercooling, indicating similarity of the transient effects for both materials. Minor influence of higher values of σ and $\Delta h/Tm\Delta h/Tm$ for indium on the relaxation time τ is predicted, while lower values of EDED and v0v0 for indium compensate each other.

Total concentration of nuclei created in a single cooling or heating run is a sum of an integral of thermal nucleation rate reduced by temperature rate (Eq. 43) and the athermal contribution dependent on temperature interval $(T0-T)(T0-T)$. The athermal contribution is independent of the cooling/heating rate (Eq. 45). Thermal nucleation provides a positive contribution to the concentration of nuclei, both for cooling and heating. Athermal nucleation, on the other side, yields positive contribution at cooling and negative one at heating.

Our model can be used to adjust processing conditions for creation desired nucleation structure. It should be noted that ideal amorphous structure free of nuclei cannot be obtained for two reasons: limited cooling rate and athermal nucleation.

ACKNOWLEDGEMENTS

This work was supported by the basic funds of the Institute of Fundamental Technological Research, Polish Academy of Sciences for the scientific research.

CONFLICT OF INTEREST

The authors declare that they have no conflict of interest.

REFERENCES

1. Frenkel J (1946) Kinetic theory of liquids. Oxford University Press, London

2. Zettlemoyer AC (ed) (1969) Nucleation. Marcel Decker, New York

3. Hoffman JD, Davis GT, Lauritzen JI Jr (1976) The rate of crystallization of linear polymers with chain folding. In: Hannay NB (ed) Treatise on solid state chemistry, vol 3. Plenum Press, New York, pp 598–600

4. Kelton KF (1991) Crystal nucleation in liquids and glasses. Solid State Phys 45:75–177

5. Kashchiev D (2000) Nucleation. Butterworth-Heinemann, Oxford

6. Kelton KF, Greer AL (2010) Nucleation in condensed matter—applications in materials and biology. Elsevier, Amsterdam

7. Kelton KF (2013) Crystal nucleation in supercooled liquid metals. Int J Microgravity Sci Appl 30:11–18

8. Volmer M, Weber E (1926) Nucleus formation in supersaturated systems. Z Phys Chem 119:277–301

9. Farkas L (1927) Keimbildungsgeschwindigkeit in übersättigten Dämpfen. Z Phys Chem (Leipzig) A125:236–242

10. Becker R, Döring W (1935) Kinetic treatment of grain formation in supersaturated vapours. Ann Physik 24:719–752

11. Zeldovich YB (1943) On the theory of new phase formation: cavitation. Acta Physicochim URSS 18:1–22

12. Turnbull D, Fisher JC (1949) Rate of nucleation in condensed systems. J Chem Phys 17:71–73

13. Eyring H (1935) The activated complex in chemical reactions. J Chem Phys 3:107–115

14. Lauritzen JI, Hoffman JD (1960) Theory of formation of polymer crystals with folded chains in dilute solution. J Res Nat Bur Std 64A:73–102

15. Hoffman JD, Lauritzen JI (1961) Crystallization of bulk polymers with chain folding: theory of growth of lamellar spherulites. J Res Nat Bur Std 65A:297–336

16. Ziabicki A (1968) Generalized theory of nucleation kinetics. I. General formulations. J Chem Phys 48:4368–4374

17. Cormia RL, Price FP, Turnbull D (1962) Kinetics of crystal nucleation in polyethylene. J Chem Phys 37:1333–1340

18. Burns JR, Turnbull D (1966) Kinetics of crystal nucleation in molten isotactic polypropylene. J Appl Phys 37:4021–4026

19. Turnbull D (1950) Formation of crystal nuclei in liquid metals. J Appl Phys 21:1022–1028

20. Turnbull D (1952) Kinetics of solidification of supercooled liquid mercury droplets. J Chem Phys 20:411–424

21. Collins FC (1955) Time lag in spontaneous nucleation due to non-steady state effects. Z Elektrochem 59:404–407

22. Kashchiev D (1969) Solution of the non-steady state problem in nucleation kinetics. Surf Sci 14:209–220

23. Chakraverty BK (1966) Kinetics of clustering process. Surf Sci 4:205–220

24. Wakeshima H (1954) Time lag in the self-nucleation. J Chem Phys 22:1614–1615

25. Feder J, Russel KC, Lothe J, Pound GM (1966) Homogeneous nucleation and growth of droplets in vapors. Adv Phys 15:111–178

26. Andres RP, Boudart M (1965) Time lag in multistate kinetics: nucleation. J Chem Phys 42:2057–2064

27. Kelton KF, Greer AL, Thompson CV (1983) Transient nucleation in condensed systems. J Chem Phys 79:6261–6276

28. James PF (1974) Kinetics of crystal nucleation in lithium silicate glasses. Phys Chem Glasses 15:95–105

29. Luborsky FE (1977) Crystallization of some Fe Ni metallic glasses. Mater Sci Eng 28:139–144

30. Scott MG (1978) The crystallization kinetics of Fe–Ni based metallic glasses. J Mater Sci 13:291–296

31. 31.

32. James PF (1985) Kinetics of crystal nucleation in silicate glasses. J Non-Cryst Solids 73:517–540

33. Kalinina AM, Filipovich VN, Fokin VM (1980) Stationary and nonstationary crystal nucleation rate in a glass of $2Na_2O$–CaO–$3SiO_2$ stoichiometric composition. J Non-Cryst Solids 38–39:723–728

34. Kelton KF, Greer AL (1988) Test of classical nucleation theory in a condensed system. Phys Rev B 38:10089Kelton KF, Greer AL (1986) Transient nucleation effects in glass formation. J Non-Cryst Solids 79:295–309

35. Kelton KF (1993) Numerical model for isothermal and non-isothermal crystallization of liquids and glasses. J Non-Cryst Solids 163:283–296

36. Kelton KF (1995) Transient nucleation in glasses. Mater Sci Eng B32:145–151

37. Shen YT, Kim TH, Gangopadhyay AK, Kelton KF (2009) Icosahedral order, frustration, and the glass transition: evidence from time-dependent nucleation and supercooled liquid structure studies. Phys Rev Lett 102:057801

38. Jarecki L (1991) Diffusion of single elements and steric limitations in kinetic theory of nucleation and crystallization. Coll Polym Sci 269:11–27

39. Jarecki L (1994) Effects of translational and rotational diffusion on the association in kinetic model of nucleation. Coll Polym Sci 272:784–796

40. Fisher JC, Hollomon JH, Turnbull D (1948) Nucleation. J Appl Phys 19:775–784

41. Ziabicki A (1968) Generalized theory of nucleation kinetics. II. Athermal nucleation involving spherical clusters. J Chem Phys 48:4374–4380

42. George HH (1985) Spinline crystallization of polyethylene terephthalate. In: Ziabicki A, Kawai H (eds) High-speed fiber spinning. Wiley, New York, pp 271–294

43. Gibbs JW (1961) The Scientific Papers of J. Willard Gibbs, vol. 1. Thermodynamics. Dover Publications Inc., New York

44. Cahn JW, Hilliard JE (1958) Free energy of nonuniform system. I. Interfacial free energy. J Chem Phys 28:258–267

45. Ziabicki A, Jarecki L (1978) Theoretical analysis of oriented and non-isothermal crystallization. III. Kinetics of crystal orientation. Colloid Polym Sci 256:332–342

46. Ziabicki A (1996) Crystallization of polymers in variable external conditions. Part 1. General equations. Colloid Polym Sci 274:209–217

47. Ziabicki A (1999) Crystallization of polymers in variable external conditions. Part 4. Isothermal crystallization in the presence of variable tensile stress or hydrostatic pressure. Colloid Polym Sci 277:752–761

48. Gutzov I, Toschev S, Marinov M, Popov E (1968) Mechanismus der Nichtstationaeren Keimbildung und des Kristallwachstums in einem Modellglase. Kristall Technik 3:337–354

49. Zhong J, Jin ZH, Lu K (2001) Melting, superheating and freezing behaviour of indium interpreted using a nucleation-and-growth model. J Phys-Condens Mat 13:11443–11452

50. Barham PJ (1984) Nucleation behavior of poly-3-hydroxybutyrate. J Mater Sci 19:3826–3834

51. Williams ML, Landel RF, Ferry JD (1955) Temperature dependence of relaxation mechanisms in amorphous polymers and other glass forming liquids. J Am Chem Soc 77:3701–3707

52. Ziabicki A, Alfonso GC (1994) Memory effects in non-isothermal crystallization. I. Theory. Colloid Polym Sci 272:1027–1042

53. Supaphol P, Spruiell JE (2000) Crystalline memory effects in isothermal crystallization of syndiotactic polypropylene. J Appl Polym Sci 75:337–346

54. Sarge SM, Höhne GWH, Cammenga HK, Eysel W, Gmelin E (2000) Temperature, heat and heat flow rate calibrations of scanning calorimeters in the cooling mode. Thermochim Acta 361:1–20

55. Barham PJ, Keller A, Otun EI, Holmes PA (1984) Crystallization and morphology of a bacterial thermoplastic: poly-3-hydroxybutyrate. J Mater Sci 19:2781–279www.webelements.com/indium. Accessed 9 Nov 2015

56. Cheng SJ, Bian XF, Zhang JX, Qin XB, Wang ZH (2003) Correlation of viscosity and structural changes of indium melt. Mater Lett 57:4191–4195

57. El-Hadi Abdel Ghaffar AM (2002) Development of a biodegradable material based on poly (3-hydroxybutyrate) PHB. PhD Dissertation, University of Halle

58. Beaucage G, Rane S, Sukumaran S, Satkowski MM, Schechtman LA, Doi Y (1997) Persistence length of isotactic poly(hydroxy butyrate). Macromolecules 30:4158–4162

INDEX